IN THE LAND OF
GREEN
LIGHTNING

IN THE LAND OF
GREEN LIGHTNING

The World of the Maya

THOR JANSON

POMEGRANATE ARTBOOKS   San Francisco

This book is dedicated to all those who love truth, and especially to Alejandra.

Published by Pomegranate Artbooks
Box 6099, Rohnert Park, California 94927

Janson, Thor, 1953–
    In the land of green lightning : the world of the Maya / Thor
Janson. — 1st. ed.
        p.   cm.
    ISBN 0-87654-079-5 : $24.95 ($35.00 Can.)
    1. Mayas. 2. Biotic communities—Mexico. 3. Biotic
communities—Central America. 4. Mexico—Description and
travel. 5. Central America—Description and travel.
    I. Title.
    F1435.J36   1994
    972'.01—dc20   94-12750
                                                        CIP

Designed by Bonnie Smetts Design
First Edition
Printed in Korea

Contents

Acknowledgments

I would like to take this opportunity to thank the following people for their invaluable help, encouragement and friendship during my sojourns in the land of the Maya: Santiago and Clara Coquix of Patchitulul, Timoteo Paz, don Juan of the Rio Tatín, doña Becky and Maestro Miguel, Professor Alvarez del Toro, Bor Mash and Rosendo Chun Pop and family. Sensei Manuel Corleto has been a constant and living reminder that only the conscious is unconscious of what the unconscious is conscious of. Professor Mario Dary Rivera, through keen interest in my work and by providing logistical support, made my initial conservation projects in Guatemala possible. Thanks go also to Juan Mario Dary, Dr. Juan de Dios Calle, Pat "the Shah of Encino" McMurray, Sumio Yukimura, Greg Neise, Caroline Misumoto, "Station Wagon" Steve, Jorge "Night Breed" Paz, Ing. Julio Piedra Santa and Carol and Mike Devine and family. Special thanks to Dr. Fernan Pavia for keeping me on my feet, to doña Patty for integrity, to Alejandra for simple wisdom and to Barry Bowen for making me feel at home in his beautiful country of Belize. Dr. Peter "the Rocker" Rockstroh provided help in the identification of unusual species and has been a major supporter in our development of programs of "eco-satire"—one lone voice laughing in the wilderness. I am indebted to Dr. Robert Carmack for being kind enough to provide his excellent contribution to the book's introductory text. To the many Mayan Indians who, over the years, have opened their hearts, shared their campfires and made me feel genuinely welcome and at home in the rain forest, I say thanks. May the Great Spirit protect and guide all of you. Finally I wish to express gratitude to Thomas F. Burke for making the publication of this book a reality.

THOR JANSON

A Brief History of the Maya

The most artistic and profoundly religious of the pre-Hispanic Mesoamerican cultures was probably the Mayan. Mayan cultural developments came quite late compared to other Mesoamerican cultures, such as those in central Mexico and Oaxaca. Nevertheless, by the beginning of the great Classic period (A.D. 200–600) the Maya had created hieroglyphic writing, calendars, counting systems, astronomical observations, stone carvings, architectural forms and painted ceramics that were unrivaled in terms of sophistication within Mesoamerica (or, for that matter, anywhere else in aboriginal America). The center of the Classic Mayan civilization was located in the lowlands of southern Mesoamerica, particularly in the Petén (Guatemala), Yucatán (Mexico), Belize, Chiapas (Mexico) and Tabasco (Mexico). Scholars have had to give up on the idea that the Classic Maya were uniquely peaceful or politically simple, but their elevated level of intellectual and artistic achievement remains unassailable.

The well-known collapse of the Classic Mayan cultures in the lowlands, beginning around A.D. 900, is still not well understood. Most scholars point to ecological pressures from intensive use of a fragile environment along with increased military pressures from outside the region. In any case, the center of cultural gravity shifted to northern Yucatán, where influence from the militaristic Toltecs of central Mexico became pervasive, and to Utatlán in Guatemala, where trade and military activities dominated social life. Scholars point out that post-Classic Mayan cultures were not less developed than their Classic period ancestors, but rather were oriented toward more secular interests.

Highland Guatemala had become the locus of the most powerful Mayan polities at the time of the arrival of the Spaniards in the sixteenth century. The Quiché-Maya kingdom, for example, reached its florescence during the fifteenth century in what is now the Department of El Quiché, Guatemala, and may have been the most powerful political system of the extensive Mayan area—and perhaps of the entire region now known as Central America. Quiché civilization was characterized by literate elites (books were painted on bark paper), large cities (some with up to 20,000 inhabitants), robust markets and commerce (trade extended from Yucatán to the north to Mexico to the west and perhaps all the way to Nicaragua to the south), complex religious cosmologies (the Quiché bible, *Popol Vuh* or "book of council," provides an example of these cosmologies) and large military organizations (armies of over 30,000 warriors could be marshaled for battle).

The Maya of what are today the countries of Guatemala, Mexico and Belize were organized as proud and independent polities when the Europeans arrived, and from the outset they resented the invasion of their world by the bearded white men from across the sea. Their conquest at the hands of the Spaniards was a profoundly traumatic historical experience, and it left a legacy of exploitation that has persisted through the centuries. Even worse, attempts to dominate the Mayan people by outsiders have been repeated in cyclical fashion since the original conquest. Nevertheless, the Maya fiercely resisted the Spaniards in 1524, and this was the basis for the other side of the Mayan historical legacy: resistance through the years to every attempt at destroying their autonomy and erasing their identity. Clearly, the Maya have been more than victims; they have been agents in a struggle to sustain a way of life in which they believe. It is truly remarkable that over six million people continue to identify themselves as Maya in the region today, and that in Guatemala they still make up over one-half of that country's overall population.

The Spanish conquest, of course, was followed by the brutal colonization of the native peoples. The Maya were forced into communities (the so-called

OPPOSITE PAGE
DAWN AT TIKAL NATIONAL PARK
GUATEMALA

"congregation" policy) in order to exploit their labor. They died of European-introduced diseases in large numbers, which, along with extensive cross-breeding (*mestizaje*), led to the virtual disappearance of Mayan populations in certain areas (Tabasco, the Petén, eastern Guatemala, the Pacific coast). On the Yucatán peninsula and in the highlands of Chiapas and Guatemala, however, the Maya closed off their communities to the Spaniards and successfully preserved much of the native culture. The Spanish response was more terror, moving the Maya to even greater distrust and hatred for their overlords.

Struggles between conservative and liberal caudillos in nineteenth-century Mexico and Guatemala brought new cycles of conquest to the Maya. In Yucatán the Maya were able to defeat their Creole overlords in battle and establish an independent state, known as the Empire of the Cross, that lasted for half a century. At midcentury in Guatemala, the conservative mestizo caudillo Rafael Carrera allowed highland Mayan communities to regain some autonomy and culturally solidify their traditional cultures. But liberal dictators—Porfirio Diaz in Mexico and Justo Rufino Barrios in Guatemala—finally ended this relative Mayan autonomy toward the end of the nineteenth century by means of a new cycle of conquest: Maya were forced to labor on henequen and coffee plantations sponsored by the despotic Mexican and Guatemalan states. The effect of the plantation regimes on the Maya was disastrous, much worse than generally realized.

Conditions improved somewhat for the Maya under the care of the revolutionary government in Mexico after the 1920s and the reform government in Guatemala between 1944 and 1954. Attempts were made to eradicate the worst features of the liberal dictatorships. The "revolutionary" governments of Mexico and Guatemala redistributed lands to the Maya and others, established fairer labor laws, permitted the founding of cooperatives and began to provide education for the native Maya and other rural peasants. These reforms have continued in Mexico, especially under the direction of the National Indigenist Institute (INI), but they were overturned in Guatemala after 1954, in large part because the wealthy plantation owners (including the United Fruit Company) were unwilling to yield any of their coercive and monopolistic power.

Conditions for the Maya of Mexico have continued to improve, although they have been under tremendous pressure to take on modern Mexican ways. In Guatemala, the long-term consequences of the U.S.-backed coup of 1954 and the counterreformation that followed have been horrendous for the Maya. This most recent cycle of conquest was precipitated by a challenge in the 1970s to the Guatemalan military government, plantation capitalists and U.S. supporters by Marxist guerrilla organizations. The guerrilla rebels actively sought the backing of the Mayan peoples of highland Guatemala and for a time succeeded in gaining widespread Mayan support. To counter the guerrillas, the Guatemalan military regime declared war against them and their Mayan supporters and in the 1970s and 1980s killed more than 100,000 of them—men, women and children. Another 1.5 million Maya sought refuge in the mountains, Guatemala City, Mexico and the U.S. The Mayan peoples who remained in the rural communities were forced into so-called "model villages" overseen by army-controlled "civil patrols." Much like the Spanish conquistadores of earlier centuries, the Guatemalan soldiers killed Maya by the tens of thousands and herded others into control communities and once again forced them to work on projects benefiting the ruling class (such as building roads in the isolated Petén zone).

Despite the cruel and sorrowful nature of Mayan history since the time of contact with the Western world, the Maya have never ceased to struggle for survival and cultural dignity, and they have achieved much in the face of adversity. As one student of Guatemalan history recently noted, "while conquest may darken their lives, it has yet to extinguish their culture." Unfortunately, peace has not yet come to many Maya, but for the first time since the original Spanish conquest almost five hundred years ago, it is now clear to almost everyone that peace cannot exist in this region of ancient Mesoamerica until justice is accorded to the Mayan people. Maya such as Rigoberta Menchú, the recipient of the Nobel Peace Prize in 1992, have made that clear, and, finally, the world has begun to understand their message.

World history is replete with illustrations of nations and groups of people who have struggled long and hard to retain their dignity and even physical existence in the face of implacable agents of the modern world, but none has fought more persistently or valiantly than the Maya. They have become the symbol of the need to preserve cultural diversity and ecological balance and of the hope for democratic change and decency throughout the Mesoamerican regions—and, indeed, the Americas as a whole.

ROBERT M. CARMACK
Director of the Center of Mesoamerican Studies
State University of New York at Albany

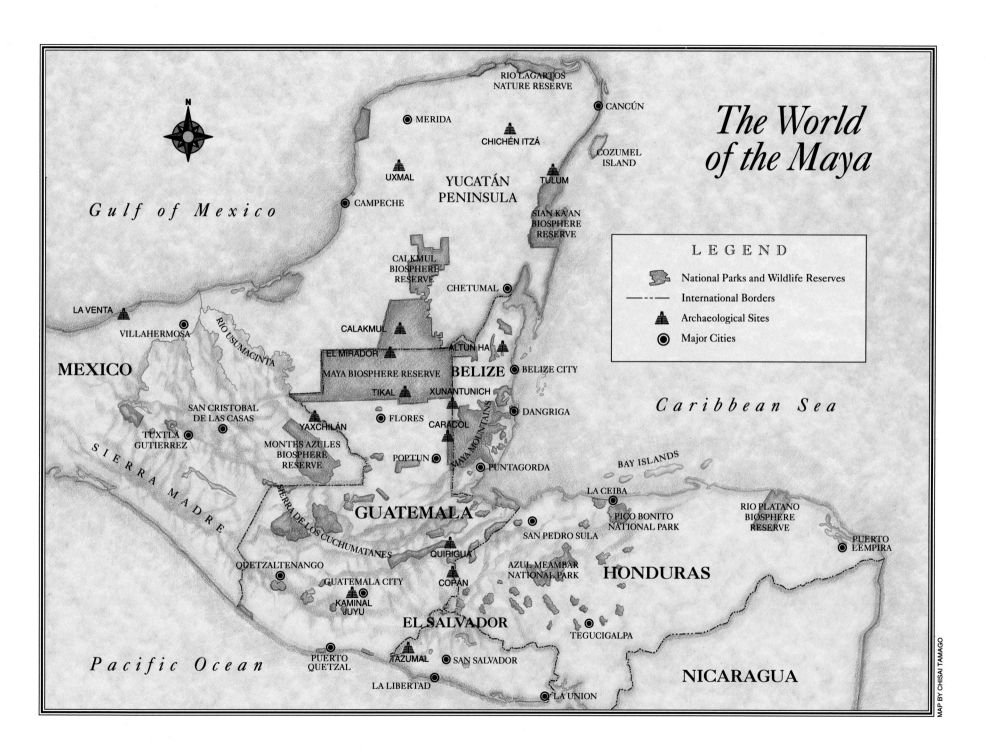

The World of the Maya

RIO LAGARTOS
NATURE RESERVE

◉ CANCÚN

◉ MERIDA

CHICHÉN ITZÁ

COZUMEL
ISLAND

▲ UXMAL

YUCATÁN
PENINSULA

▲ TULUM

◉ CAMPECHE

Gulf of Mexico

SIAN KA'AN
BIOSPHERE
RESERVE

CALKMUL
BIOSPHERE
RESERVE

CHETUMAL ◉

LEGEND

National Parks and Wildlife Reserves

International Borders

▲ Archaeological Sites

◉ Major Cities

LA VENTA ▲

VILLAHERMOSA ◉

RIO USUMACINTA

CALAKMUL ▲

ALTUN HA ▲

MEXICO

EL MIRADOR ▲

MAYA BIOSPHERE RESERVE

BELIZE

◉ BELIZE CITY

SAN CRISTOBAL
DE LAS CASAS ◉

TIKAL ▲

XUNANTUNICH ▲

◉ DANGRIGA

Caribbean Sea

◉ FLORES

TUXTLA
GUTIERREZ ◉

YAXCHILÁN ▲

CARACOL ▲

MONTES AZULES
BIOSPHERE
RESERVE

SIERRA DE LOS CUCHUMATANES

POPTUN ◉

MAYA MOUNTAINS

◉ PUNTAGORDA

BAY ISLANDS

S
I
E
R
R
A

M
A
D
R
E

LA CEIBA ◉

RIO PLATANO
BIOSPHERE
RESERVE

GUATEMALA

PICO BONITO
NATIONAL PARK

PUERTO
LEMPIRA ◉

QUIRIGUA ▲

SAN PEDRO SULA ◉

AZUL MEAMBAR
NATIONAL PARK

HONDURAS

QUETZALTENANGO ◉

GUATEMALA CITY ◉

COPÁN ▲

KAMINAL
JUYU ▲

Pacific Ocean

EL SALVADOR

TEGUCIGALPA ◉

PUERTO
QUETZAL ◉

TAZUMAL ▲

◉ SAN SALVADOR

NICARAGUA

LA LIBERTAD ◉

◉ LA UNION

The Natural World

 For centuries the land of the Maya has been of particular interest to biologists because of its unique location. It is the crossroads of two of Earth's major life zones: the Nearctic Realm (North America) and the Neotropical Realm (South America). It is here, in the bio-geo-graphical heart of the continent, where many Neotropical plant and animal species reach their northern limits of distribution, while large numbers of Nearctic genera and species reach their southern range of dispersal. Maples and oaks grow alongside mahoganies and zapotes; coyotes and mountain lions share the forest with olingos, howler monkeys and jaguars.

Over vast periods of time, and at least four ice ages, major migrations of both plants and animals occurred across the narrow isthmus. As glacial ice melted, the oceans rose, inundating much of Central America. Many species became isolated on mountaintops that had become, temporarily, islands. Today the ancient tropical highland forests are home to large numbers of unique species and have the highest rate of endemism.

European naturalists who mounted the first expeditions to the American rain forest were astounded by the incredible exuberance and variety of life they found. Alfred Russel Wallace, who is credited along with Charles Darwin for developing the theory of natural selection, recorded his observations of the area in 1895:

Here no one who has any feeling of the magnificent and the sublime can be disappointed; the sombre shade, scarce illuminated by a single direct ray even of the tropical sun, the enormous size and height of the trees, most of which rise like huge columns a hundred feet or more without throwing out a single branch, the strange buttresses around the base of some, the spiny or furrowed stems of others, the curious and even extraordinary creepers and climbers which wind around them, hanging in long festoons from branch to branch, sometimes curling and twisting on the ground like great serpents, then mounting to the very tops of the trees, thence throwing down roots and fibres which hang waving in the air, or twisting around each other form ropes and cables of every variety and size and often of the most perfect regularity. These, and many other novel features—the parasitic plants growing on the trunks and branches, the wonderful variety of the foliage, the strange fruits and seeds that lie rotting on the ground—taken altogether surpass description, and produce feelings in the beholder of admiration and awe. It is here, too, that the rarest birds, the most lovely insects, and the most interesting mammals and reptiles are to be found. Here lurk the jaguar and the boa-constrictor, and here amid the densest shade the bell-bird tolls his peal.

The plant life in the land of the Maya includes approximately eight thousand species of vascular plants due to the region's vast diversity of conditions: climates vary from very hot in desert areas to cold in the mountains, and the soil reflects a diverse geological history, from ancient mountains to relatively recent volcanoes.

If one approaches the Yucatán peninsula from the eastern sea, the first ecosystem encountered is the coral reef. Running virtually unbroken from Cancún down to Amatique Bay, this is considered the second-largest barrier reef in the world. The reef itself is the result of tiny animals related to jellyfish building little shelters of calcium carbonate, one on top of another. After millions of years, these structures have become today's magnificent reef. The reef ecosystem is in many ways the marine equivalent of the rain forest

OPPOSITE PAGE
IZALCO VOLCANO

Cerro Verde National Park, El Salvador
The sunset casts a red hue over Izalco Volcano. In the distance you can see the Pacific coastline through the mist. Though most of El Salvador has been deforested due to overpopulation and the effects of a decade-long war, an excellent system of national parks, including Cerro Verde, has miraculously survived.

(both are considered Earth's most productive life systems), with its seemingly endless array of tiny, brightly hued tropical fish, squadrons of fierce-looking barracuda, meandering parrot fish and groupers, hair-raising hammerhead sharks, an occasional gentle manatee and, moving inland, a shallow bottom covered in turtle grass where large queen conch roam.

The next life zone is the mangrove swamp, which covers many of the islands and most of the coastline. These dense, bug-infested forests provide nesting grounds for many bird species, nurseries for fish and shelters for a wide variety of animals. Further inland beyond the coastal mangroves looms the lofty canopy of the rain forest. Huge trees reach up toward the sun, their trunks whitened with lichens and covered with vines, making them very difficult to identify. Within the forest large "islands" of savanna, typified by coarse sandy soil, support a variety of grasses and bushy plants and small stands of slash pine.

Heading up into the central highlands is a series of inland plateaus and valleys where pronounced wet and dry seasons and moderate temperatures favor forests of pine, juniper and cypress and a profusion of flower-ing herbaceous plants. Higher up into the high sierra coniferous forests predominate, and above 10,000 feet they give way to alpine meadows where tussock grasses, succulents and giant agaves abound. Along certain mountain ridges and volcanic peaks lies the mysterious cloud forest. Made up of a large variety of broad-leaved trees and nearly constantly enshrouded in fog and mist, this is the home of the resplendent quetzal, the emerald toucanet and the howler monkey.

Along the volcanic slopes that lie ten to thirty miles inland from the Pacific coast exists a forest type called bocacosta, which is composed of tree ferns, palms and large trees covered with vines and epiphytes. Inland from the Pacific mangrove swamp is a forest with characteristics unlike those of any other in the region. The land is flat and covered with giant silk-cotton trees. Mixed among them are many other tree species, including sweet-pea, golden-yellow buttercup and spectacular yellow tabebuias.

All these habitats provide refuge for a growing number of endangered species such as jaguars, tapirs, kinkajous, trogons, eyelash vipers, harpy eagles and more than one hundred species of North American migratory birds. Sadly, these tropical ecosystems are increasingly endangered themselves due to unwise and selfish exploitation by the human inhabitants. Laws have been passed during the last ten years that provide legal protection to extensive tracts of forest in the Maya, Calakmul and Sierra de las Minas Biosphere Reserves. But these reserves are nothing more than "paper parks" as long as corrupt officials ignore the activities of illegal logging operations.

When I first visited Central America in 1973, the area that is now the Maya Biosphere Reserve was 98 percent forested. According to the most recent satellite data only 35 percent of that forest remains. This remnant will be destroyed in the next ten years unless something is done to stop the transnational logging mafia. Big money is at stake, so only international pressure can force local governments to uphold laws and provide the protection necessary to save what is the largest remaining block of rain forest north of the Amazon basin. Readers are encouraged to support organizations dedicated to the defense of planetary ecology.

THOR JANSON

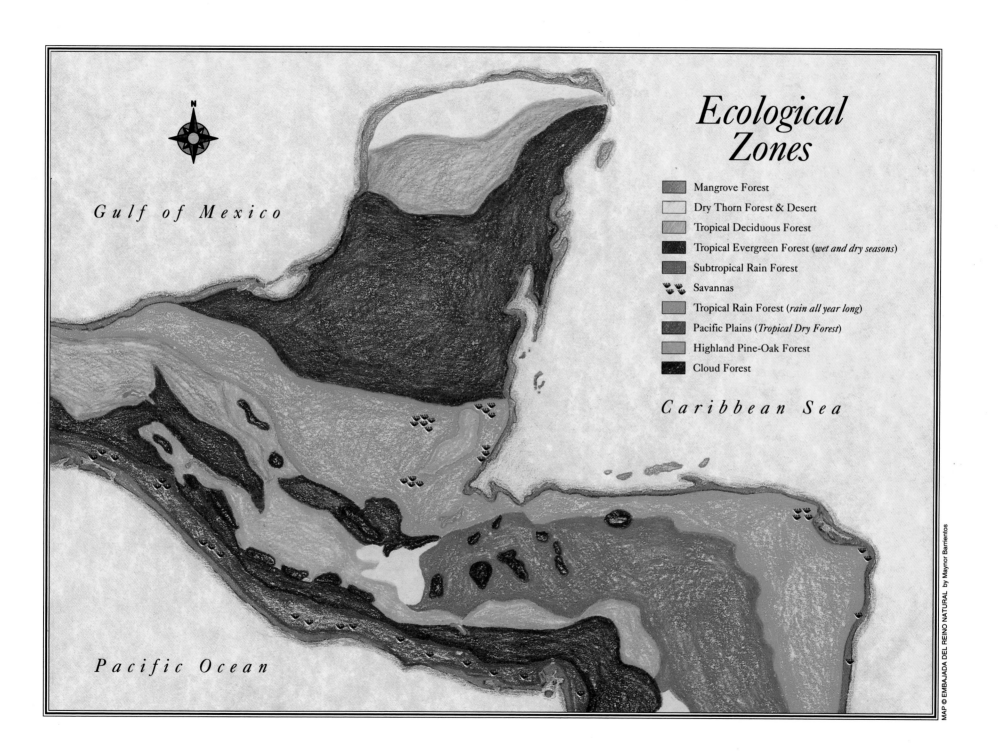

Ecological Zones

Legend:
- Mangrove Forest
- Dry Thorn Forest & Desert
- Tropical Deciduous Forest
- Tropical Evergreen Forest (*wet and dry seasons*)
- Subtropical Rain Forest
- Savannas
- Tropical Rain Forest (*rain all year long*)
- Pacific Plains (*Tropical Dry Forest*)
- Highland Pine-Oak Forest
- Cloud Forest

Gulf of Mexico

Caribbean Sea

Pacific Ocean

IN THE LAND OF
GREEN
LIGHTNING

Of emeralds and rubies you were formed
The Jewel of the Cloud Forest
A shimmering bolt of Green Lightning
The Resplendent Quetzal-serpent.

Ancient sages named you Kukulcan
Their supreme symbol of Light and Freedom
The Heart of Heaven, the Herald of Tatixel
Huahop, Owner of the world, giver of Wisdom.

PREVIOUS PAGE
RESPLENDENT QUETZAL *Pharomachrus mocinno*

SIERRA YALIHÚX, ALTA VERAPAZ PROVINCE, GUATEMALA

Long before the first people invaded Central America, Kukul, the quetzal, was undisputed lord of the cloud forest. The male quetzal's spectacular emerald green and ruby red colors and meter-long serpentine tail so inspired the Maya and the Aztecs that they named their central god-man deities after him: Kukulcan and Quetzalcoatl. Immense pyramids were erected to honor the winged serpent, and it was a capital offense to kill the bird. Today, in contrast, the quetzal is in danger of extinction.

This spectacular male was photographed while he was feeding in an aguacatillo tree (*Lauraceae* sp.), which is a close relative of the common avocado and is one of the quetzal's favorite foods.

LEFT
SCARLET MACAW *Ara macao*

MAYA BIOSPHERE RESERVE, EL PETÉN PROVINCE, GUATEMALA

It was once a common sight to see pairs of these magnificent birds, which inhabited the jungle up to an elevation of nine hundred meters, flying just above the forest canopy, squawking raucously. Although the Maya have traditionally exploited this bird as a source of food and for its crimson, blue and yellow feathers for ceremonial costumes, it was not until recently that these macaws were ruthlessly hunted down to be sold around the world as pets. Today they have become virtually extinct throughout their former range.

OPPOSITE
OCELLATED TURKEY *Meleagris ocellata*

TIKAL NATIONAL PARK, GUATEMALA

This brightly colored turkey is found exclusively in the low-elevation forests of the Yucatán peninsula, Belize and northern Guatemala and has become rare because of over-hunting. At

OPPOSITE
PINK FLAMINGO *Phoenicopterus ruber*
RÍO LAGARTOS NATIONAL PARK, YUCATÁN, MEXICO

RIGHT, TOP
Feathers of the RESPLENDENT QUETZAL
Pharomachrus mocinno
SIERRA YALIHÚX, ALTA VERAPAZ PROVINCE, GUATEMALA
Clouds of mystery surround the enigmatic figure of the winged
serpent, the resplendent quetzal. Chosen among all creatures as
the supreme symbol of light and life by the Aztecs, Toltecs and
Maya, the images of Quetzalcoatl-Kukulcan adorn nearly all of
the thousands of pyramids and related monuments of
Mesoamerica.

RIGHT, BOTTOM
Feathers of the SCARLET MACAW *Ara Macao*
MAYA BIOSPHERE RESERVE, GUATEMALA

VAMPIRE BAT mother and young

CERRO HUECO NATURE RESERVE, CHIAPAS, MEXICO

COATIMUNDI *Nasua narica*

SIERRA DE LAS MINAS BIOSPHERE RESERVE, GUATEMALA

These small mammals are closely related to the raccoon and are fairly common in the undisturbed rain forest. Once while visiting Tikal I saw a troop of at least fifty individuals cautiously crossing a road in perfect single file, older ones first followed by juveniles and then babies. All held their tails erect and looked as if they were on a grade school outing.

KINKAJOU *Potos Flavos*

MONTES AZULES BIOSPHERE RESERVE, CHIAPAS, MEXICO

These small monkeylike omnivores live in the rain forest canopy. Extremely agile and largely nocturnal, kinkajous love sweet foods and are known locally as *ositos mieleros,* "little honey bears."

IN THE LAND OF GREEN LIGHTNING

LEFT
BOBCAT *Lynx rufus*
CHIAPAS, MEXICO
Reaching the limits of their range in southern Mexico, bobcats are now extremely rare in the remnant forests of the border area between Chiapas and Oaxaca.

OPPOSITE
MARGAY *Felis Weidii*
CALAKMUL BIOSPHERE RESERVE, CAMPECHE, MEXICO
Monkeylike climbing ability distinguishes this rare cat as the most arboreal of the Mesoamerican felines. Spending much of its time in the rain forest canopy, the margay stalks small mammals and birds, leaping from limb to limb and hanging from branches for extended periods with the help of its semiprehensile tail. Habitat destruction and over-hunting for its beautiful fur have brought the margay close to extinction.

IN THE LAND OF GREEN LIGHTNING

OPPOSITE

JAGUAR *Felis onca*

COCKSCOMB BASIN, STANN CREEK DISTRICT, BELIZE

The largest and most spectacular cat in the Americas, the jaguar is one of the most commonly seen animal figures in the pre-Columbian art of Mesoamerica. Known to the Maya as "Balam" and to the Aztecs as "Tezcatlipoca," the jaguar symbolized the power of death. He was the lord of the underworld in constant struggle with the winged serpent, the lord of light.

RIGHT

JAGUARUNDI *Felis yaguarundi*

RÍO BRAVO CONSERVATION AREA, ORANGE WALK DISTRICT, BELIZE

Known locally by the Mayan name "Ekmuch," the jaguarundi has also been called the "otter cat" because of its unusual appearance, red-brown coloration and affinity for water. Jaguarundi are excellent swimmers and rarely climb trees as they search for small mammals, birds and reptiles.

GOLDEN EYELASH VIPERS *Bothrops schlegelii*

CELAQUE NATIONAL PARK, HONDURAS

One of several arboreal pit vipers to inhabit the region, the eyelash viper, although relatively small (it rarely exceeds half a meter in length), is considered to have a very active venom. Because these vipers live in trees and bushes, chances are that if one strikes a human it will likely hit the head, neck or forearm, increasing the severity of the toxic reaction.

FER-DE-LANCE *Bothrops asper*

RÍO DULCE NATIONAL PARK, IZABAL PROVINCE, GUATEMALA

Without question the most feared viper in northern Central America, the fer-de-lance is also known as "la barba amarilla" or "yellow beard" due to the coloration of its lips and throat. This viper sometimes reaches nearly ten feet in length and is the cause of most snakebite-related deaths in the region.

BLACK-COLLARED HAWK *Busarellus nigricollis*

MAYA BIOSPHERE RESERVE, GUATEMALA

Also known as the "fishing hawk," this medium-sized bird of prey ranges from Mexico down to Brazil. It is considered quite rare and is most often seen around marshes and swamps in the most remote sections of the lowland rain forest.

ORNATE HAWK-EAGLE *Spizaetus ornatus*

TIKAL NATIONAL PARK, GUATEMALA

This large, striking bird is distinguished by its big black crest and bright cinnamon face and neck. It is still relatively common in the virgin lowland rain forests of Central America. Its prey consists of small mammals, reptiles, birds and insects.

RED-TAILED HAWK *Buteo jamaicensis*

LAKE ATITLÁN NATIONAL PARK, SOLOLÁ PROVINCE, GUATEMALA

This large bird is the most common soaring hawk found in the mountains of Guatemala. Some are permanent residents while others are migrating to Central America in the winter from western North America. They prey primarily on small rodents.

LEFT

TAPIR *Tapirus bairdii*

CALAKMUL BIOSPHERE RESERVE, CAMPECHE, MEXICO

The tapir, known locally as the "danta," is the largest terrestrial herbivore in the Americas. Adults reach a length of nearly six feet and weigh seven hundred pounds or more. Over the millennia tapirs have established well-defined paths during foraging missions. These same paths came to be used by humans as the easiest way to traverse the jungle. Today, numerous highways have been built along the original tapir routes.

OPPOSITE

ARMADILLOS *Dasypus novemcinctus*

COCKSCOMB BASIN WILDLIFE SANCTUARY, STANN CREEK DISTRICT, BELIZE

These little mammals are so abundant in southern Belize that scientists were surprised to find that they are the primary prey of the jaguar in this region. The Maya call them "iboy," and they are prized as a source of meat. They live in burrows and feed mostly on insects.

IN THE LAND OF GREEN LIGHTNING

SPIDER MONKEY *Ateles geoffroyi*

MAYA BIOSPHERE RESERVE, GUATEMALA

Extremely agile, these medium-sized monkeys can be seen leaping through the canopy from limb to limb in exuberant feats of acrobatics. Their fully prehensile tail serves as a fifth hand, and they are often seen hanging upside down, sustained only by their tail and one hind paw, tranquilly munching a savory fruit or nut.

CAPUCHIN MONKEY *Cebus capucinus*

RÍO PLATANO BIOSPHERE RESERVE, HONDURAS

Also known as white-faced monkeys, the capuchins are highly social canopy dwellers often seen foraging in bands of ten or fifteen individuals. Sometimes, being intensely curious beasts, they come right down to the edge of a jungle camp to get a better look at their strange two-legged cousins.

HOWLER MONKEY *Alouatta villosa*

COMMUNITY BABOON SANCTUARY, BERMUDIAN LANDING, BELIZE

Heard from a distance the howler's amazing vocalizations conjure up images of some large, ferocious beast. These highly social monkeys live in troops, often with as many as twenty individuals, and are constantly on the move searching the forest canopy for succulent leaves, fruits and nuts.

HORNED GUAN *Oreophasis derbianus*

TECPÁN RIDGE, CHIMALTENANGO PROVINCE, GUATEMALA
The horned guan is one of the rarest and most unusual birds in the world. Found exclusively in the cloud forests of a few volcanic peaks and windswept mountain ridges in Chiapas and Guatemala, this turkey-sized guan is recognized instantly by the large, strawberry-colored horn on its head. Its call is an eerie mooing. It is close to becoming extinct because of the destruction of its habitat.

OPPOSITE
KEEL-BILLED TOUCAN *Ramphastos sulfuratus*

RÍO DULCE NATIONAL PARK, IZABAL PROVINCE, GUATEMALA
The largest of three species of toucans found in northern Central America, this beautiful bird is easily identified by its bright yellow breast and huge multicolored bill. Found in lowland jungle up to an elevation of two thousand feet, the keel-billed toucan spends much of its day searching for food in the form of fruits, small insects and reptiles, and by robbing other birds' nests of eggs. Its call is a froglike "grrik-grrik-grrik." Captured in large numbers to be sold into the pet trade, this toucan is increasingly hard to find outside of reserve areas.

IN THE LAND OF GREEN LIGHTNING

OPPOSITE

LIZARD *Sceloporus malachiticus*

CERRO HUECO NATURE RESERVE, CHIAPAS, MEXICO

These small, completely harmless lizards are among the most commonly seen reptiles in the rain forest.

RIGHT

BANDED BASILISK LIZARD *Bassiliscus vittatus*

MONTES AZULES BIOSPHERE RESERVE, CHIAPAS, MEXICO

This little dinosaur reaches a length of half a meter, though most of this is made up by its long slender tail. It is known in some regions as the "Jesus Christ lizard" because of its amazing ability to run at high speed on top of pools of water.

BELOW

GREAT CURRISAW *Crax rubra*

CALAKMUL BIOSPHERE RESERVE, CAMPECHE, MEXICO

This very large and robust bird, with a long tail and prominent crest of erectile, forward-curled feathers, spends its days in the forest looking for fallen fruits and insects. Great currisaws are one of the most prized food animals among the lowland Maya.

GREEN EYELASH VIPER *Bothrops schlegelii*
BIOTOPO MARIO DARY RIVERA QUETZAL SANCTUARY,
BAJA VERAPAZ PROVINCE, GUATEMALA
This is the green color phase of the eyelash viper, which also
occurs in a brilliant yellow. Excellent camouflage renders this
viper virtually invisible in the foliage.

BELOW
TROPICAL RATTLESNAKE *Crotalus durissus*
LAKE ATITLÁN NATIONAL PARK, SOLOLÁ PROVINCE, GUATEMALA
Along with the jaguar and the quetzal, the rattlesnake is ubiquitous in Mayan art. The ever-present figure of the feathered serpent is composed of elements of the rattlesnake and quetzal and
is represented in some form on almost all Mayan pyramids.
Reaching six feet in length, this viper should be approached
with great caution.

OPPOSITE
SPECTACLED CAIMAN *Caiman crocodilus*
RÍO SARSTUN, IZABAL PROVINCE, GUATEMALA
The smallest of the three representatives of the order crocodilia
found in northern Central America, the caiman inhabits the
rivers and swamps of coastal lowlands. Exploitation of their
beautiful skins has caused them to become quite rare.

BEADED LIZARD *Heloderma horridum*

MOTAGUA VALLEY, ZACAPA PROVINCE, GUATEMALA

The only venomous lizard in the region, the beaded lizard or heloderm inhabits arid lowlands. Heloderms are not equipped with injecting fangs—as vipers are—but have simple venom glands that secrete the potent liquid into small grooves in their teeth. For this reason, when the lizard bites its victim, it must hold on and chew the wound in order to disperse the venom. Once a heloderm bites, it is almost impossible to get it to release its grasp without breaking its jaw. Fortunately, although common in desert areas, these lizards are sedate and non-aggressive.

OPPOSITE

BUSHMASTER *Lachesis mutus*

RÍO PLATANO BIOSPHERE RESERVE, HONDURAS

This is the giant among pit vipers, often reaching lengths in excess of twelve feet. Its Latin name translates to "silent fate," so named for its ability to strike without audible warning—an ability that makes the bushmaster extremely dangerous.

IN THE LAND OF GREEN LIGHTNING

GREEN IGUANID LIZARD *Laemactus longipes*

PALENQUE NATIONAL PARK, CHIAPAS, MEXICO

Due to this reptile's excellent camouflage it is very difficult to detect in the foliage. These lizards are so confident about their invisibility that they will allow a person to almost touch them before they run away.

RIGHT

TREE FROG *Smilisca baudinii*

CHAN CHICH NATURE RESERVE, ORANGE WALK DISTRICT, BELIZE

Equipped with suction cups for finger pads, allowing them to leap with incredible agility from leaf to leaf, these small amphibians hunt for small insects in their rain forest canopy home.

KATYDID

Cerro Cahuí Reserve, El Petén Province, Guatemala

Cryptic coloration and physical shape create near perfect camouflage for this insect: it becomes invisible when sitting among leaves. Katydids' familiar chirping is one of the ubiquitous sounds of the rain forest.

RIGHT
IO MOTH Larva *Automenis* sp.

Biotopo Mario Dary Rivera Quetzal Sanctuary, Alta Verapaz Province, Guatemala

OPPOSITE
MOTH

Cuchumatanes Mountains, Huehuetenango Province, Guatemala

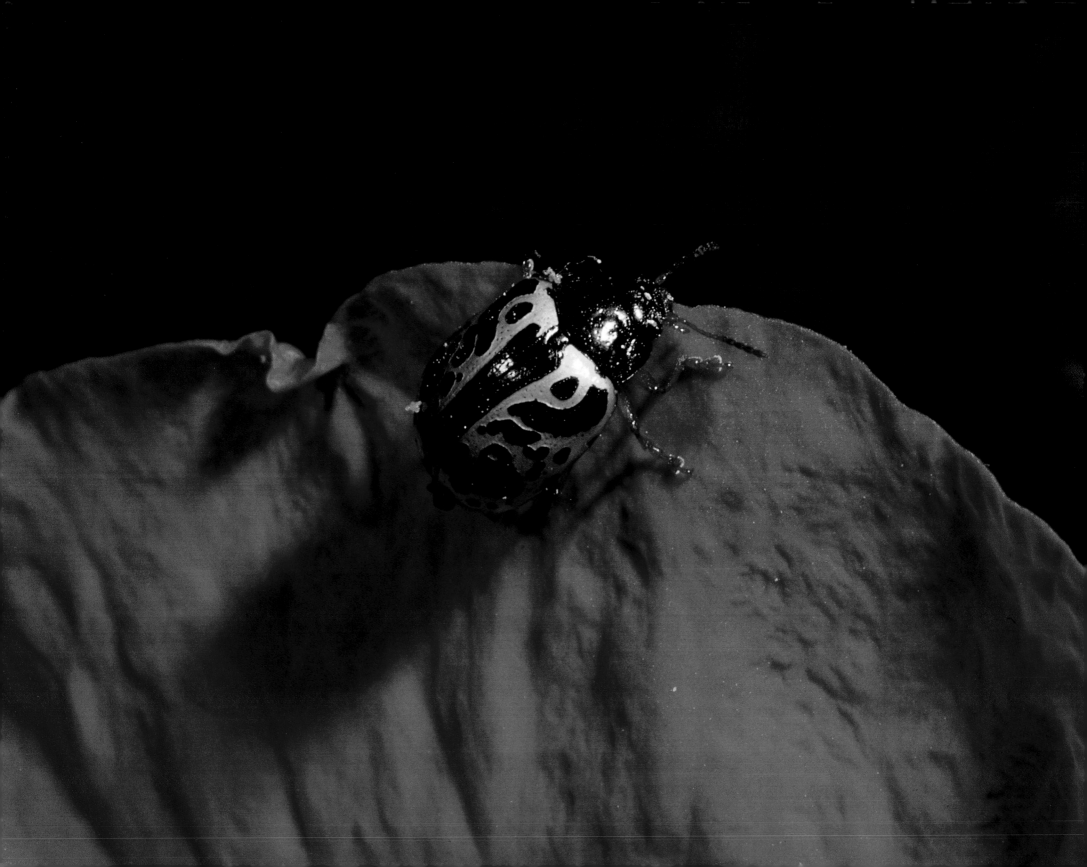

OPPOSITE
BEETLE genus *Chrysomlidae*
FUEGO VOLCANO NATIONAL PARK, GUATEMALA

RIGHT
BUTTERFLY *Battus polydamas*
SHIPSTERN NATURE RESERVE, COROZAL DISTRICT, BELIZE

IN THE LAND OF GREEN LIGHTNING

OPPOSITE
BUTTERFLY *Papilio polyxenes*
CHAN CHICH NATURE RESERVE, ORANGE WALK DISTRICT,
BELIZE

RIGHT
BUTTERFLY *Eucides isabella*
SHIPSTERN NATURE RESERVE, COROZAL DISTRICT, BELIZE

BELOW
BUTTERFLY *Euptoieta hegesia*
RÍO BRAVO CONSERVATION AREA, ORANGE WALK DISTRICT,
BELIZE

IN THE LAND OF GREEN LIGHTNING

LEFT
ORCHID
Laguna Belgica Nature Reserve, Chiapas, Mexico

BELOW
WILDFLOWER
Cobán Region, Alta Verapaz Province, Guatemala

OPPOSITE, LEFT
HIBISCUS FLOWER *Hibiscus rosa-sinensis*
Maya Mountains Forest Reserve, Cayo District, Belize

OPPOSITE, RIGHT
CACTUS FLOWER
Lake Atitlán National Park, Sololá Province, Guatemala

WILDFLOWER

Maya Mountains Forest Reserve, Belize

SUCCULENT FLOWER

Cuchumatanes Mountains, Huehuetenango Province, Guatemala

WILDFLOWER with MINIATURE CRAB SPIDER

Macal River Valley, Belize

IN THE LAND OF GREEN LIGHTNING

OPPOSITE
CICADA

LAGUNA BELGICA NATURE RESERVE, CHIAPAS, MEXICO

RIGHT
MANTIS

LAGUNAS DE MONTEBELLO NATIONAL PARK, CHIAPAS, MEXICO

IN THE LAND OF GREEN LIGHTNING

LA CATARÁTA DE PANAJACHEL

Sololá Province, Guatemala

THOUSAND-FOOT FALLS

Mountain Pine Ridge Forest Reserve, Cayo District, Belize

Also known as "Hidden Valley Falls," this long, slender plume of water is actually over 1,600 feet long, making it the highest waterfall in Central America. The valley below is virtually unexplored and remains the domain of Balam, the jaguar.

CELAQUE NATIONAL PARK

Honduras

This enchanted cloud forest, usually enshrouded in mist, is home to the resplendent quetzal and the emerald toucanet. These are the most ancient forests in Central America, where undisturbed evolution has slowly been evolving unique species for millions of years.

LEFT
KEK'CHI Indian boys in the Cloud Forest
SIERRA YALIHÚX, ALTA VERAPAZ PROVINCE, GUATEMALA

BELOW
LA TIGRA NATIONAL PARK
HONDURAS

OPPOSITE
BUTTRESSED TREE
GUANCASTE NATIONAL PARK, BELIZE

IN THE LAND OF GREEN LIGHTNING

SIAN KA'AN BIOSPHERE RESERVE

QUINTANA ROO, MEXICO

Sian Ka'an was established by presidential decree in 1986 for the purpose of protecting 528,000 hectares of prime coastal habitat. The reserve encompasses vast expanses of tropical evergreen forest, mangrove swamp and coral reef ecosystems.

TOP

BACALAR LAGOON

QUINTANA ROO, MEXICO

BOTTOM

MONTERICO WILDLIFE RESERVE

SANTA ROSA PROVINCE, GUATEMALA

This reserve, located on the Pacific coast, protects mangrove swamps and salt water marshes that provide shelter to more than one hundred species of resident and migratory birds and serve as hatcheries for large numbers of marine and fresh water fish.

THE BALSAM COAST
LIBERTAD PROVINCE, EL SALVADOR

Traditionally, this beautiful section of Pacific coastline was important for its production of a fragrant reddish brown liquid extracted from the *Myroxylon pereirae* tree. This balsam extract is still used today in a variety of medicines and cosmetics and is valued for its soothing and tonic effects upon the body.

BLACK CARIB village
PUNTA SAL NATIONAL PARK, HONDURAS

TZUTUHÍL Indian boy

LAKE ATITLÁN NATIONAL PARK, GUATEMALA

Every morning boys from the town of Santiago Atitlán paddle out on the lake in their cayucos to fish. These sturdy canoes, fashioned from large tree trunks, are their main form of transportation.

BOTTOM

Native fisherman

SIAN KA'AN BIOSPHERE RESERVE, QUINTANA ROO, MEXICO

OPPOSITE

PLACENCIA TOWN

STANN CREEK DISTRICT, BELIZE

Placencia is a quiet little fishing village located at the tip of a narrow peninsula that separates pristine Placencia Lagoon from the Caribbean Sea. On the left is the combined post office and telephone exchange.

LEFT
RÍO TULIJA
CHIAPAS, MEXICO

OPPOSITE
PALO DE FUEGO (Fire Tree)
QUINTANA ROO, MEXICO

IN THE LAND OF GREEN LIGHTNING

RÍO LAGARTOS NATIONAL PARK

YUCATÁN, MEXICO

Located on the northern edge of the Yucatán peninsula, Río Lagartos protects large expanses of salt marsh, mangrove and dune ecologies affording refuge to many endangered species.

RIGHT

MANATEE *Trichechus manatus*

HICKS CAY, BELIZE

The largest aquatic herbivore in Central America, the manatee had an original range along the entire Atlantic coast. Mayan fishermen prepared dried manatee meat for a food they called "bucan." With the European invasion of America, many of the pirates and freebooters came to rely upon bucan as a staple food, and they became known as "buccaneers." Today manatees are extremely rare and have become extinct in most of their former territory.

IN THE LAND OF GREEN LIGHTNING

RED LEAFY CORAL *Wellsopphyllia* sp.

PUNTA ALLEN, QUINTANA ROO, MEXICO

Most types of coral, including this one, emerge to feed only at night.

RAINBOW PARROT FISH *Scarus guacamaia*

LIGHTHOUSE REEF, BELIZE

MAM Indian dwelling
Chuy Village, Huehuetenango Province, Guatemala

Village of CHIUL
El Quiché Province, Guatemala

MAM Indian dwelling

TODOS SANTOS CUCHUMATÁN, HUEHUETENANGO PROVINCE,
GUATEMALA

OPPOSITE, LEFT

Village of TODOS SANTOS CUCHUMATÁN

HUEHUETENANGO PROVINCE, GUATEMALA

OPPOSITE, RIGHT

LOS CUCHUMATANES NATIONAL PARK

HUEHUETENANGO PROVINCE, GUATEMALA

One of the highest mountain ranges in Central America, the Cuchumatanes are unique in the land of the Maya. Thousand-foot sheer rock escarpments face ridges wooded with tropical pine forests. Higher up above the tree line beautiful mist-enshrouded alpine meadows are marked by huge agaves. Here it is too high and cold for cultivating corn, so the Indians dedicate themselves to shepherding.

CENOTE XKEKEN

YUCATÁN, MEXICO

Cenotes are created when the roof of a cavern caves in, forming a sink hole. The Itza-Maya of the Yucatán have always considered them sacred, and this is understandable in a region of arid limestone plains where there are virtually no rivers. These cool, subterranean pools have been their main source of water.

TOP

LOLTÚN CAVERNS

YUCATÁN, MEXICO

These hand prints are thought to date from somewhere around 200 B.C., though no one knows who made them. The record of early human migration and settlement remains obscure.

BOTTOM

LOLTÚN CAVERNS

YUCATÁN, MEXICO

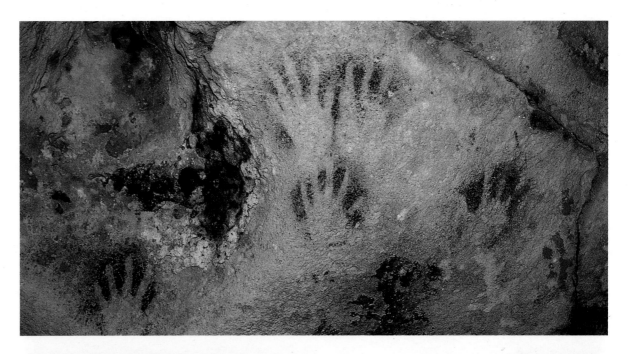

Main highway through the CUCHUMATANES MOUNTAINS

HUEHUETENANGO PROVINCE, GUATEMALA

CAÑON DEL SUMADERO NATIONAL PARK

CHIAPAS, MEXICO

From the rim of this tremendous canyon the visitor is given a spectacular view of the Santo Domingo River more than three thousand feet below. According to legend, hundreds of Indian warriors were surrounded here by the Spanish conquistadores, but rather than submit they hurled themselves into the abyss.

LAKE ATITLÁN NATIONAL PARK

Sololá Province, Guatemala

Located in the very heart of the Mayan realm, Atitlán has been one of the most important sacred places since humans arrived here some ten thousand years ago. Shamans claim that the crystal blue waters are charged with natural forces which, when bathed in, can cure disease. To the south are the towering conical peaks of Toliman and Atitlán volcanoes.

BOTTOM

LAKE ATITLÁN NATIONAL PARK

Sololá Province, Guatemala

More than a mile deep, Atitlán's opal-like emerald green, turquoise blue color constantly changes during the day. Nestled along the shore are twelve Cak'Chiquel and Tzutuhíl Indian villages, where the inhabitants continue to live much as they did one thousand years ago.

OPPOSITE

LAKE ATITLÁN NATIONAL PARK

Sololá Province, Guatemala

IN THE LAND OF GREEN LIGHTNING

SALCAJÁ VALLEY

QUETZALTENANGO PROVINCE, GUATEMALA

The Cak'chiquel Indian farmers who live here owe their prosperity to rich volcanic soils ideal for vegetable cultivation. Santa Maria Volcano towers in the background.

RIGHT

LOS CUCHUMATANES NATIONAL PARK

HUEHUETENANGO PROVINCE, GUATEMALA

One of the highest roads in Central America, Route 9-N weaves across this enchanted plateau past the villages of Paquix and Cinabal on the way to the regional center of Ixtatán.

BELOW

ORCHID

SIERRA YALIHÚX RESERVE, ALTA VERAPAZ PROVINCE, GUATEMALA

LEFT

Welcome to Guatemala

EL PETÉN PROVINCE, GUATEMALA

Part of the army public relations campaign to terrorize the citizenry, signs like this appear near military bases.

BELOW

MAYA MOUNTAINS FOREST RESERVE

CAYO DISTRICT, BELIZE

I came upon this sign one day while hiking through the forest. Only later did I learn that I had stumbled upon a British military training area. The British maintain troops stationed in Belize to protect the tiny country from possible Guatemalan invasion.

OPPOSITE

Village of SAN JUAN IXCOY

HUEHUETENANGO PROVINCE, GUATEMALA

The valley where San Juan Ixcoy is situated is an important apple-growing center. Here cool springtime weather predominates year round.

LEFT

Market Day, TODOS SANTOS CUCHUMATÁN

HUEHUETENANGO PROVINCE, GUATEMALA

OPPOSITE, LEFT

CAK'CHIQUEL Indians

LAKE ATITLÁN NATIONAL PARK, GUATEMALA

OPPPOSITE, RIGHT

Don Juan of the Rio Tatín

IZABAL PROVINCE, GUATEMALA

Self-described as being "part Indian, part Spaniard, part black and part pirate," don Juan lives on a small river in the jungle near the Caribbean coast. An expert bushman, he can survive quite nicely almost anywhere with an ax, a machete and a sack of corn.

LEFT
CAK'CHIQUEL Indian Girl
Santiago Sacatepéquez, Guatemala

BELOW
LACANDÓN Indian Boy
Lacanjá, Chiapas, Mexico

OPPOSITE, LEFT
CAK'CHIQUEL Indian Man
Patchitulul, Sololá Province, Guatemala
Eighty-seven-year-old Santiago Coquix shucks corn from the annual harvest. For the Maya every aspect of the production of corn—from planting according to the indications of the calendar priest to the final grinding for tortillas, tamales and the fermented drink koosha—is carried out with ceremony and prayers of thanksgiving.

OPPOSITE, RIGHT
MAM Indian Girl
Todos Santos Cucchumatán, Huehuetenango Province, Guatemala

The Chicken Bus

TODOS SANTOS CUCHUMATÁN, HUEHUETENANGO PROVINCE, GUATEMALA

Dubbed "chicken buses" by amused travelers, these vehicles are the main form of transport linking remote Indian villages with the outside world. The experience of sharing one's seat with a few chickens or a piglet and a bus full of splendidly dressed and exceedingly friendly Indians is a magical experience not to be forgotten.

RIGHT

Five hundred years of exploitation

SOLOLÁ, SOLOLÁ PROVINCE, GUATEMALA

"1492–1992: 500 Years of Exploitation, Discrimination, and Repression of the Mayan People Fighting for Their Rights and Liberty!" This protest took place in the regional capital to commemorate the European "discovery" of the "new world." Only a few years ago this sort of public protest would have been quickly and brutally dispersed by the military. Today the voices crying for freedom and self-determination for native inhabitants cannot be silenced.

Market Day
NEBAJ, EL QUICHÉ PROVINCE, GUATEMALA

BELOW
Little MAM Indian girl dressed in ceremonial costume
TODOS SANTOS CUCHUMATÁN, HUEHUETENANGO PROVINCE, GUATEMALA

OPPOSITE
Market
CHICHICASTENANGO, EL QUICHÉ PROVINCE, GUATEMALA

IN THE LAND OF GREEN LIGHTNING

OPPOSITE

Catholic church at TOTONICAPÁN

TOTONICAPÁN PROVINCE, GUATEMALA

RIGHT

ANTIGUA

SACATEPÉQUEZ PROVINCE, GUATEMALA

In its day Antigua was one of the greatest cities in the Spanish
colonial empire, rivaling Lima and Mexico City in importance. It
was the capital of the "Auciencia de Guatemala," which included
all of Central America and Chiapas. Several earthquakes deci-
mated the city in 1773, and the capital was relocated to the site of
today's Guatemala City.

BELOW

Colonial chapel

ALTA VERAPAZ PROVINCE, GUATEMALA

This detail of an image painted in oil on stucco adorns the altar
of an abandoned chapel on a coffee plantation near the town of
San Juan Chamelco.

OPPOSITE
Giant Spirit Kite

CHIMALTENANGO PROVINCE, GUATEMALA

Every year on All Saints Day the villagers of Santiago Sacatepéquez fly spectacular giant kites that have taken them weeks or even months to build. The kites can reach more than ten feet in diameter and are flown from the cemetery. Believers say that the ritual helps them communicate with departed souls.

RIGHT
Holy Week Procession

CHICHICASTENANGO, EL QUICHÉ PROVINCE, GUATEMALA

Chichicastenango is a Nahuatl word meaning the "Place of the Nettles." Here ancient Mayan religious beliefs meld with Catholic rituals, creating a curious hybrid faith.

IN THE LAND OF GREEN LIGHTNING

OPPOSITE
Holy Week Procession

ANTIGUA, SACATEPÉQUEZ PROVINCE, GUATEMALA

Ancient Mayan rituals blend with Catholic tradition to create the most colorful Easter processions in the Americas.

RIGHT
Holy Week Procession

ANTIGUA, SACATEPÉQUEZ PROVINCE, GUATEMALA

On Thursday night the streets are carpeted with vibrantly colored paintings made from tinted sawdust. Early the next morning massive platforms are borne on the shoulders of penitents reenacting Christ's walk to the cross.

LEFT

Children dressed in ceremonial costume

CHICHICASTENANGO, EL QUICHÉ PROVINCE, GUATEMALA

OPPOSITE

Ceremonial dancers

TZALAMILÁ, ALTA VERAPAZ PROVINCE, GUATEMALA

Once a year in this remote village located at the edge of the Yalihúx cloud forest, native men and boys participate in dances dressed up in brilliant costumes representing deer, jaguar and monkey personages and white bearded conquistadores. These dances, which have their origin in the ancient past, re-create major historical events as well as aspects of the Mayan supernatural world.

BAILE DE LA CONQUISTA

SAN JUAN LA LAGUNA, SOLOLÁ PROVINCE, GUATEMALA

Also known as the "Dance of the Moors," the Baile de la Conquista is performed once a year in each Indian village. The dance is always accompanied by the ceaseless drone of the marimba, bass drum and an oboelike instrument called the "chilimia," and it may commence in the morning or afternoon, perhaps continuing all night. It is considered a great privilege for a man or boy to be selected by committee to be one of the dancers. Among the characters represented in the dance are Pedro de Alvarado, the infamous Spanish conquistador who ordered many Indian chiefs crucified and initiated nearly five hundred years of slavery and oppression of the Maya. He is accompanied by several Spanish ladies, also played by Indian men. Tecun Uman, the great Mayan chief slain on the battlefield of Xelaju by Alvarado, is accompanied by a witch doctor or "brujo" as well as several jaguars, deer and dogs. The dogs are usually played by small boys and are the clowns of the dance, sneaking up to nip Alvarado or performing ridiculous pantomimes just out of his grasp. While the performance is not choreographed in any discernible way, there are favorite themes that are enacted over and over throughout the day, such as a jaguar carrying away a Spanish woman and two deer sparring. The more imaginative or unusual the skit, the greater are the crowds' shouts of approval. While it is impossible for a non-Indian to fully appreciate the meaning these dances hold for the natives, one thing is clear: in the Baile de la Conquista it is the Indians who win and the Spanish who are tormented and vanquished.

IN THE LAND OF GREEN LIGHTNING

OPPOSITE
Supplicants bring offerings to Maximón

ZUNIL PUEBLO, QUETZALTENANGO PROVINCE, GUATEMALA

The cult of Maximón is widespread all over the Mayan highlands. Thousands of Indians claim to have been cured of all manner of illness by the benevolence of this enigmatic mustached figure. But who is Maximón? Some say he is Judas; others say he is an evil Indian saint.

LEFT
MAXIMÓN, Lord of ATITLÁN

SANTIAGO ATITLÁN, SOLOLÁ PROVINCE, GUATEMALA

There are three major Maximón figures in Guatemala located in Zunil, San Andres Itzapa and Santiago Atitlán, as well as many minor ones. The centuries-old conflict between the Christian church and traditional shamanism continues, and the cult of Maximón, with its thousands of devotees, is at the forefront of this spiritual war.

RIGHT
MAXIMÓN OF SAN ANDRÉS ITZAPA

CHIMALTENANGO PROVINCE, GUATEMALA

OPPOSITE

TEMPLE OF THE WARRIORS

CHICHÉN ITZÁ RUINS, YUCATÁN, MEXICO

Chichén Itzá was founded in the year A.D. 432 and became an important population center. In the tenth century it was taken over by invading Toltecs, who were responsible for some of its larger buildings, including this temple. The cornice is decorated with warriors, eagles and jaguars devouring human hearts.

RIGHT

TEMPLE OF KUKULCAN

CHICHÉN ITZÁ RUINS, YUCATÁN, MEXICO

The heads of the feathered serpents on either side of the stairway represent the god Kukulcan, the herald of light and life.

BELOW

Detail of stela

COPÁN RUINS, HONDURAS

Of all the inhabitants of the pre-Columbian new world, only the Maya had a fully developed script and were able to write whatever they could speak. Stelas were used to record important historical events in the lives of the Mayan rulers.

COLOSSAL HEAD

LA VENTA ARCHAELOGICAL MUSEUM, TABASCO, MEXICO

Weighing nearly twenty tons, this enormous sculpture was created by artisans of the Olmec culture, which was in full flower by 1200 B.C. The consensus among archaeologists in 1993 was that these faces represent "were-jaguars," the melding of a human with his animal nagual. Nevertheless, the decidedly African-looking visage has led some scholars to speculate that the Olmec represented a race distinct from other Mesoamerican peoples.

OPPOSITE, LEFT

Detail of stela

COPÁN RUINS, HONDURAS

Monoliths were erected by the Classic Maya elite to bear testimony of themselves and leave a record of their life and times. Only during the last decade have Mayan scholars been able to unravel the meaning held in the ancient hieroglyphics engraved in stone more than one thousand years ago.

OPPOSITE, RIGHT

CEIBAL RUINS

EL PETÉN PROVINCE, GUATEMALA

A relatively minor site during the Classic period, it is thought that Ceibal grew in importance after being invaded by colonists from the north who brought with them a distinctly non-Mayan architectural style, including round platforms associated with the cult of Quetzalcoatl. The ancient Maya employed elaborate carvings on stone stelas, such as the one pictured here, to record such occurrences as royal coronations and important military victories, as well as important historical events.

PYRAMID OF THE SOOTHSAYER
UXMAL RUINS, YUCATÁN, MEXICO

This is also known as the "Pyramid of the Wizard" from a legend that states it was built by the son of a witch, who was born from an egg. The structure is actually made up of five different levels built in distinct periods and representing different styles.

PALACE OF THE GOVERNOR
UXMAL RUINS, YUCATÁN, MEXICO

This site was first settled in the seventh century. In the tenth century the Xiues, arriving from the central plateau, took over the city and imposed the worship of the gods Tláloc and Quetzalcoatl.

IN THE LAND OF GREEN LIGHTNING

PALACE OF THE GOVERNOR

UXMAL RUINS, YUCATÁN, MEXICO

Uxmal, the home of the Xiu tribe, was founded in 1007, but its finest structures, such as the one shown here, were built a century later. The statue in the foreground represents two jaguars joined at the breast, pointing north and south.

Woman holding mustached figure

LA VENTA ARCHAEOLOGICAL MUSEUM, TABASCO, MEXICO

Some scholars are convinced that Phoenician sailing ships had reached both the east and west coasts of Central America by 250 B.C. The male figure appearing in this carving, dated c. 200 B.C., does not resemble a Mesoamerican Indian.

Funerary urn

EL PETÉN PROVINCE, GUATEMALA

When members of the Mayan aristocarcy died, they were often placed in ceramic containers before being buried in stone crypts. This beautiful example is adorned with stylized jaguar figures and skulls. It dates from the late Classic period, around the year A.D. 800.

CHAC MOOL figure

CHICHÉN ITZÁ RUINS, YUCATÁN, MEXICO

At the entrance to the Temple of the Warriors rests this enigmatic reclining figure. These Chac Mool figures are thought to have been used as sacrificial altars and reflect the influence of Toltec culture, which was centered to the north near Mexico City.

Anthropomorphic figure

COPÁN NATIONAL PARK, HONDURAS

The cult of the monkey-man was widespread all over Mesoamerica. The monkey was deified as patron of artists, musicians and scribes, just as the Egyptians had made the baboon-god Thoth patron of their scribes. Copán reached its height as one of the most important of ancient Mayan cities in the sixth and seventh centuries. The city was abandoned shortly after A.D. 820.

THE JAGUAR THRONE, TEMPLE OF KUKULCAN

CHICHÉN ITZÁ RUINS, YUCATÁN, MEXICO

Painted red and encrusted with jade, this artifact is located in a secret chamber, apparently an altar room, deep within the pyramid.

TEMPLE OF THE GIANT JAGUAR

TIKAL NATIONAL PARK, GUATEMALA

Towering more than forty-four meters above the Great Plaza, this pyramid is the hallmark of Tikal. Deep inside archaeologists discovered the tomb of Ah Cacau ("King Chocolate"), who had ascended Tikal's throne in A.D. 682. The temple's high roof comb provides a superb vantage point for observing jungle life.

LEFT

TULUM RUINS NATIONAL PARK

Quintana Roo, Mexico

The jungle meets the crystal blue waters of the Caribbean at the ancient city of Tulum. Although there is evidence of human occupation since the sixth century, it was not until around 1200 that Tulum achieved prominence as one of the most important centers on the east coast of Yucatán.

OPPOSITE

TULUM RUINS NATIONAL PARK

Quintana Roo, Mexico

IN THE LAND OF GREEN LIGHTNING

OPPOSITE

TIKAL NATIONAL PARK

EL PETÉN PROVINCE, GUATEMALA

Considered by many to be the most magnificent of all Mayan sites, this national park was established to protect the integrity of extensive ruins, including some of the region's greatest pyramids. The park also contains nearly four hundred square kilometers of rain forest, creating the region's premier wildlife reserve, where jaguars, howler monkeys and innumerable other species find refuge from humanity's onslaught.

RIGHT

CARACOL ASTRONOMICAL OBSERVATORY

CHICHÉN ITZÁ RUINS, YUCATÁN, MEXICO

Caracol Observatory is so named because it was built in the form of a snail. From here scientist-priests made highly accurate observations and predictions concerning the movements of heavenly bodies at a time when Europe was groping through the Dark Ages. Small windows in the structure are perfectly aligned to follow the rising and setting sun at the spring and autumn equinoxes.

PACAYA VOLCANO ERUPTION

PACAYA VOLCANO NATIONAL PARK, GUATEMALA

On a beautiful, crisp January day in the highlands I drove up to the slope of Pacaya Volcano with my girlfriend, Max. I heard it was erupting, and I wanted to get some good shots for my files. By four o'clock Pacaya was spewing molten lava several hundred meters into the air every thirty to forty-five seconds. The frequency and intensity continued to increase; by five o'clock the earthshaking explosions were reaching an unusual and rather frightening level of ferocity. Villagers from the nearby town of Caracol began a hasty evacuation, passing by us on the road, some on foot, some in pickups. The people, poor Indian campesinos, were terrified. Some were praying out loud, beseeching God to have mercy on them; women were weeping and children were crying. Major tremors rocked the earth, and Max and I were beginning to feel the heat. She was ready to leave, but I was too excited at the prospect of getting some great photographs, so she agreed to wait a little longer. By six p.m. everything had intensified. Villagers, now hysterical, were running down the road; men were yelling at us to abandon the area or be killed. Pacaya was beginning to resemble a gigantic Roman candle. Every ten seconds the earth rumbled and quaked and large quantities of lava spewed forth, creating an incandescent river flowing down the southeast slope. Refusing to wait any longer, Max jumped into the car, threw it into reverse and, unfortunately, launched it backwards off the road and into powderfine black ash. The car was stuck; our means of escape suddenly had vanished. I made a futile attempt to jack the car out, but continued snapping pictures, much to Max's fury. I thought we would have to run for it, but everything was changing so fast, and in spite of the danger I was reluctant to leave. The eruptions were almost continuous, and a strange, deep, ultra-low-frequency sound vibration was emanating from the ground, from the air—all around us. The solid column of lava was now spewing perhaps half a kilometer into the sky, lighting up the entire landscape with a fiery orange red glow. Just when I was ready to pick up my camera bag and start running down the slope, a pickup loaded with stragglers came speeding down the road. Seeing our predicament, several men jumped out, tied a cable on the car and helped Max to free it. During all this I stayed at the camera while the column of lava got higher and higher and the horrendous omnipresent vibration shook the world. A few seconds later I was sitting in the passenger seat, rocketing down the switchbacks leading to the valley below. I could not help but be transfixed at the scene out of the rear window: the whole sky was alight, and a brilliant fire cloud, like an enormous plume, stood above the volcanic cone and reached at least five kilometers up into the upper atmosphere. Lightning bolts shot through the clouds. We were lucky to be alive.

Several days later we learned that most of the rock and ash had fallen south and had not destroyed any of the villages, though many roofs of houses had caved in. Only two human fatalities were reported. More than two hundred cattle and horses in the pastures on the south slope were killed. The plume of fiery ash had shot up ten kilometers into the stratosphere, sending ash into the jet stream. The column of lava we saw had reached a height of nearly two kilometers. The eruption was one of the largest to be reported during this century in Central America.

IN THE LAND OF GREEN LIGHTNING

OPPOSITE
LAKE ATITLÁN NATIONAL PARK
SOLOLÁ PROVINCE, GUATEMALA

RIGHT
Moonlight vista over the POLOCHIC VALLEY
ALTA VERAPAZ PROVINCE, GUATEMALA

This splendid view was obtained from the crest of the Sierra Yalihúx mountains at an elevation of nearly nine thousand feet. In the distance, clouds cascade down the slopes of the Piedras Blancas mountains.

IN THE LAND OF GREEN LIGHTNING

Male QUETZAL at nest site

SIERRA YALIHÚX, ALTA VERAPAZ PROVINCE, GUATEMALA

Quetzals make their nests in rotten tree trunks, and both the male and female incubate and feed the young. Here the male is bringing food to the chicks. If you look closely you can see a superbly camouflaged chick at the door of the nest hole.

STANN CREEK DISTRICT, BELIZE

Each year thousands of acres of rain forest are put to the torch to make room for cattle ranches and orange groves, so that these products can be sold for a few pennies less in North America and Europe.

**EMBAJADA
DEL REINO NATURAL**

EMBAJADA DEL REINO NATURAL

The present decade has become a race against time for people who are involved in trying to save the earth's remaining tropical rain forests. This is especially true for Central America, where most of the forests have already been destroyed. Establishing nature reserves is essential, but unless there is popular support among the citizenry no long-term conservation effort will be successful. Effective environmental education is a crucial element in the fight to defend nature. Embajada del Reino Natural (The Embassy of Nature) is an international organization dedicated to producing "green" educational materials to be used in schools and to impact the public at large. We must unite in the protection and defense of the planet's natural ecology if human life is to continue aboard Spaceship Earth. Those interested in helping our group or obtaining further information may contact Thor Janson, c/o Pomegranate Artbooks, Box 6099, Rohnert Park, California 94927 U.S.A.

Exploring the World of the Maya

 The listings of archaeological sites, national parks and wildlife reserves and Mayan settlements on the following pages are mere samplings of what the archaeology buff and adventurous naturalist will encounter while traveling the Mayan route. There are literally hundreds of ruins sites and nature reserves in the region. Some are small, such as the Cerro Huitepec cloud forest in the mountains of southern Mexico; some are huge, like the Maya Biosphere Reserve in Guatemala. The reader is advised to keep in mind that park development in the land of the Maya is still in its infancy, and many of the sites and reserves have little or no infrastructure or on-site management. If you are looking for well-marked trails, guided tours and snack bars, then keep to the better-known places such as Chichén Itzá and Tulum in Mexico's Yucatán peninsula or the Biotopo del Quetzal in Guatemala. If wilderness trekking appeals to you, then head out to the Maya Mountains in Belize or El Triunfo National Park in Mexico's Sierra Madre. Luxury accommodations are generally available only in the larger cities and resorts adjacent to the major archaeological sites. Outside of these areas hotels range from modest to "ultra-budget."

Major Archaeological Sites

Mexico

LA VENTA, VILLAHERMOSA, TABASCO

Notable among this amazing collection of stelas and stone carvings are the colossal stone heads of the Olmec artisans. The Olmecs suddenly appeared in 1200 B.C. and are thought to have been the people who gave rise to the Mayan civilization.

PALENQUE NATIONAL PARK, CHIAPAS

Established as a national park in 1981, this site preserves 1,771 hectares of rain forest. Palenque, one of the most important of all Mayan sites, was a major ceremonial center in A.D. 615, when Lord Pacal assumed the throne at age twelve. Under his rule and the later reigns of his sons, great temples were built exhibiting some of the most unusual architectural styles in the entire Mayan world. The site sits on the side of the Tumbala Hills, which rise suddenly from the flat Atlantic plains.

BONAMPAK NATURAL MONUMENT, CHIAPAS

Located in the primeval Lacandón Jungle, this site of 4,357 hectares can be reached by vehicle via a deeply rutted dirt track, a drive of several hours from Palenque. Bonampak was built between A.D. 200 and 900 and is deservedly famous for the brilliantly colored frescoes inside one of its temples. These are considered to be the best-preserved examples of Mayan painting anywhere.

TONINÁ RUINS, CHIAPAS

Toniná is located near the Tzeltal Indian town of Ocosingo and is presently being excavated. Some very unusual carvings on the temples have been uncovered. This is an excellent place to see archaeologists at work.

YAXCHILÁN NATURAL MONUMENT, CHIAPAS

This park protects 2,621 hectares of rain forest. Because it is located on the bank of the Usumacinta River, Yaxchilán is one of the most picturesque and exotic of all Mayan sites. It is still being used as a ceremonial center by the Lacandón Indians. It reached its height of splendor in the Classic period and was abandoned around A.D. 900. This is one of the more remote sites, reached by several hours of dirt track and then by canoe, starting from Palenque.

EDZNÁ RUINS, CAMPECHE

The name "Edzná" means "House of Grimaces" and refers to the stone masks originally located on the roof comb of the huge Temple of Five Levels. This temple is unique in that evidence suggests that its four lower levels were used as residences of priests and nobility. The temple overlooks the enormous central plaza, three times the size of a football field. Edzná is located near the state capital of Campeche.

RIO BEC RUINS, CAMPECHE

Located in the south-central portion of the Yucatán peninsula, this site provides an excellent opportunity to visit the remains of a major Mayan city—which is still almost completely undisturbed by archaeologists. The site can be reached by a four-wheel-drive vehicle from the town of Xpujil.

CALAKMUL RUINS, CAMPECHE

One of the most massive pyramids in the whole Mayan world (its base covers two hectares) is situated fifty-five kilometers southwest of Rio Bec in the heart of the Calakmul Biosphere Reserve. The site is located at the edge of a beautiful lagoon and is surrounded by a seemingly endless sea of jungle.

XPUJIL RUINS, CAMPECHE

The unusual triple-tower structure of this ancient ceremonial site can be seen from the highway that runs between Escárcega to Chetumal. It lies on the outskirts of the town Xpujil, which has been continuously occupied since A.D. 400.

HORMIGUERO RUINS, CAMPECHE

Also near the town of Xpujil and recently opened to the public, these partially excavated ruins exhibit a refined architectural style. Of note are details on the frieze of the main temple, which may indicate the existence of a phallic cult.

TULUM RUINS, QUINTANA ROO

A post-Classic fortress and trading center located on the Caribbean coast, the buildings at Tulum date from A.D. 1200. This was the only Mayan city still occupied when the Spanish arrived in 1518. "Tulum" means "City of the New Dawn," which alludes to the spectacular sunrises over the ocean that bathe the city in fiery color.

COBÁ RUINS, QUINTANA ROO

Covering an area of eighty square miles, Cobá was one of the greatest of all Mayan cities. Some 20,000 structures have been identified, and estimates indicate that more than 40,000 people lived here during the city's height (A.D. 600–900). Less than 5 percent of the site has been excavated.

CHICHÉN ITZÁ, YUCATÁN

The most completely restored of all Mayan sites, Chichén Itzá was founded in A.D. 415 by the Itzá-Maya people who migrated from Cozumel Island. After about two hundred years the site was abandoned when the Itzá moved south into the Petén rain forest. Three hundred years later their descendants returned to restore Chichén Itzá to its former glory. As one of the three city-states forming the League of Mayapan, it reigned supreme in the Yucatán until 1204, when the alliance was destroyed by civil war.

UXMAL RUINS, YUCATÁN

Founded in the seventh century, Uxmal is considered to be among the finest examples of Mayan architecture in the Yucatán. Among its outstanding structures is the unique oval-shaped Pyramid of the Magician.

MAYAPÁN RUINS, YUCATÁN

Mayapán was the last post-Classic capital of the Yucatán. For more than two centuries the three-way alliance between Chichén Itzá, Mayapán and Uxmal overshadowed the entire region. This site is rarely visited by tourists today because it is largely overgrown with jungle.

Belize

LAMINAI RUINS, ORANGE WALK DISTRICT

Located on the banks of the New River Lagoon, Laminai is one of the most impressive sites in Belize. Though it does not compare in size or splendor to the major sites in Mexico or Guatemala, Laminai is fascinating in that it looks much the same today as it did when first discovered by European explorers. Of the sixty or so structures that have been identified, most are still uncovered. The 385-hectare archaeological reserve also provides protection for the surrounding jungle.

ALTUN HA RUINS, ORANGE WALK DISTRICT

This site was occupied between 1,000 B.C. until about A.D. 900. Large quantities of jade and obsidian trade objects were found here, although neither jade nor obsidian occur naturally in Belize, indicating that Altun Ha was an important trade center. The core of the site includes several medium-sized pyramids and temples.

Xunantunich Ruins, Cayo District

Xunantunich is one of the only Mayan cities to be built on a hill. The largest pyramid at the site, known as "El Castillo," is second in height only to the recently discovered great pyramid at Caracol. These two structures continue to be the largest buildings in Belize. Xunantunich flourished during the Classic period; there is evidence that a major earthquake largely destroyed the city around A.D. 900 and may have been the cause of its abandonment.

Caracol Ruins, Cayo District

Caracol was discovered in 1936 by a gang of "chicleros" as they explored the jungle in search of the raw material used to produce chewing gum. But it was not until 1985 that the first detailed archaeological analysis of the site revealed the city's true stature: Caracol rivaled even Tikal in size and importance. Its largest pyramid, "Canaa," reaches forty-two meters in height. Carvings on altars tell of wars between Caracol and Tikal, with victories alternating between the two.

Lubaantún Ruins, Toledo District

"Lubaantún" means "Place of the Fallen Stones," an apt description of this largely unexcavated site located in the dense rain forest west of Punta Gorda Town. All of the buildings here were assembled without the use of mortar, each stone being painstakingly cut to exactly fit another. Evidence suggests that the site was occupied for only a short time between A.D. 700 and 890.

Guatemala

Tikal National Park, El Petén Province

Ranking among the world's greatest ancient ruins, Tikal encompasses 575 square kilometers of nature reserve. It was one of the first areas of rain forest to be set aside for the protection of endangered animals such as the jaguar. At the end of the sixth century the city of Tikal became the region's primary center of trade, science and religion. The population rose to include some 100,000 people, and the city itself extended to cover an area of fifty square kilometers.

Uaxactún Ruins, El Petén Province

Twenty-four kilometers north of Tikal's main plaza lie the largely unrestored ruins of Uaxactún, Tikal's rival city. Beneath one of Uaxactún's temples lie the remains of an earlier structure, thought to date from 2000 B.C., making it the oldest building ever found in the Petén.

Yaxjá Ruins, El Petén Province

Yaxjá is almost completely unrestored and is situated at the edge of a clear blue lake of the same name. The town seems to have been laid out on a grid pattern, unlike most other Mayan centers where urban development was less systematic. A small island in the middle of Lake Yaxjá is the site of the ruins of Topoxte.

El Mirador Ruins, El Petén Province

Rivaling or even surpassing Tikal in size and importance, El Mirador can be reached either by hiking two days north from the little town of Carmelita or by helicopter. Several massive pyramid structures rise above the forest canopy, and the largest of these, the Dante Complex, rises in three levels to a height of seventy meters, making it the tallest known Mayan structure.

Río Azul Ruins, El Petén Province

Río Azul is another remote site, somewhat more accessible than El Mirador. A dirt track, usually not transitable even by four-wheel-drive vehicles, links Río Azul with Uaxactún and Tikal and provides an excellent route for a three- or four-day trek. The site sits adjacent to the pure waters of the Blue River, populated by an abundance of crocodiles. Several splendid tombs were uncovered here, lined with white plaster with figures painted in bright red. Unfortunately, large amounts of artifacts were carried off by looters before the Guatemalan government arranged to post guards at the site.

Ceibal Ruins, El Petén Province

Surrounded by pristine rain forest this remote site can be reached either by boat or jeep from the frontier town of Sayaxché on the bank of the Río de la Pasión. The site has been partially cleared and restored, and a series of trails leads to a variety of stone altars, stelas and ceremonial monuments, all visited by incomparably huge swarms of mosquitoes.

Quiriguá Ruins, Izabal Province

Quiriguá is best known for its collection of magnificent and enormous stelas located around the entire extent of its Great Plaza. The nine major stelas in the plaza are the tallest in the Mayan world (the largest is eight meters high and weighs sixty-five tons) and exhibit some of the best examples of Mayan carving.

KAMINALJUYÚ RUINS, GUATEMALA PROVINCE

Located in the northwestern suburbs of Guatemala City, this site is one of the oldest Mayan ruins. Originally built around 300 B.C. by farming people known as the Miraflores, Kaminaljuyú at its height had a population of 50,000, making it one of the largest Mayan cities. Around A.D. 300 the city was conquered by Toltec invaders, and a fascinating synthesis of Mayan and Toltec art can be seen at the site.

IXIMCHÉ RUINS, TECPÁN PROVINCE

Iximché, five kilometers south of the town of Tecpán, was built as a fortress and capital of the Cak'Chiquel Maya. Founded in 1470, its reign was short lived: Pedro de Alvarado arrived in 1524 and had the Indian chiefs crucifed and burned at the stake. The city, too, was put to the torch, and its surviving inhabitants fled into the refuge of the surrounding forest. Iximché is one of the best examples of post-Classic highland Mayan architecture.

ZACULEU RUINS, HUEHUETENANGO PROVINCE

Zaculeu, located near the town of Huehuetenango, is another highland fortress and ceremonial center. Unfortunately, it was reconstructed in 1947 by the United Fruit Company in what has been called one of the worst restoration jobs in the history of archaeology. The temples and buildings were completely covered in cement and painted white. However, Zaculeu remains important as one of the few remaining examples of post-Classic Mayan culture.

Honduras

COPÁN RUINS, COPÁN PROVINCE

Along with Tikal and Chichén Itzá, Copán is ranked as one of the most important of all Mayan sites. Human settlers first arrived here around 1000 B.C., but Copán did not emerge as a major city until the sixth century. The ruins include some of the finest stone carving to be found anywhere in the Mayan world.

El Salvador

TAZUMUL RUINS, SANTA ANA PROVINCE

Some archaeologists claim that this site was first occupied nearly seven thousand years ago. The buildings visible today, including a fourteen-step pyramid, were built by the Pipil Indians around A.D. 980.

National Parks and Wildlife Reserves

Mexico

EL TRIUNFO BIOSPHERE RESERVE, CHIAPAS

This enormous park has been established largely thanks to the tireless efforts of Mexico's pioneering conservationist Miguel Alvarez del Toro. The 119,177-hectare park includes both slopes of the Sierra Madre and ranges from 500 to 2,850 meters in elevation, protecting large expanses of pristine cloud forest, mountain rain forest and pine-oak-liquidambar forest. This is one of the most important reserves in the American hemisphere and provides refuge for countless species including the endangered quetzal, horned guan and jaguar. The park is managed by the Instituto de Historia Natural of Chiapas, and information is available from their offices located at the zoo in Tuxtla Gutiérrez.

LAGUNA BELGICA NATURE RESERVE

Located a short drive from the state capital of Tuxtla Gutiérrez, this small reserve is Mexico's first "educational park." Comprising just over forty-seven hectares, the area was set aside for students to observe the processes of nature as they exist without human interference. The reserve includes tropical evergreen forest and pine-oak-liquidambar forest and surrounds a small lagoon and swamp.

LAGOS DE MONTEBELLO NATIONAL PARK

This park of over six thousand hectares includes some sixty-eight lakes and lagoons. Vegetation types include tropical deciduous forest, evergreen forest and cloud for-est. Lagos de Montebello is home to the resplendent quetzal, the margay and many beautiful orchid varieties.

CAÑON DEL SUMIDERO NATIONAL PARK, CHIAPAS

Over a vast period of time the Grijalva River, which runs along a geological fault line, eroded this enormous canyon (its walls reach 1,200 meters in height). When the Chiapa Indians were pursued by Spanish invaders, they decided to throw themselves into this canyon rather than surrender to or be slaughtered by the conquistadores. The park includes about 28,000 hectares and protects lowland jungle up to pine-oak forest at elevations reaching 1,720 meters.

CERRO HUITEPÉC NATURE RESERVE, CHIAPAS

Located halfway along the road between San Cristobal de las Casas and San Juan Chamula, this small reserve protects one of the last cloud forests in the region. Administered by the private conservation organization PRONATURA, the park offers a good system of trails where visitors may catch a glimpse of a mountain trogon.

RANCHO NUEVO ECOLOGICAL PARK, CHIAPAS

Rancho Nuevo, located a few kilometers outside of the town of San Cristobal, includes a series of mountain ridges at an elevation between 2,300 and 2,700 meters. In this pine-oak forest the keen-eyed natural-ist may see trogons, flying squirrels, owls and several species of viper. The park also is known for its caves, which are lit with electric lamps for a distance of one kilometer and are loaded with beautiful formations.

LA ENCRUCIJADA ECOLOGICAL RESERVE, CHIAPAS

This huge reserve runs along the Pacific coast for over 140 kilometers, from Chocohuital Estuary in Pijijipan to Cubildos Lagoon near the Guatemalan border. It includes salt marsh, savanna, zapoton-mangrove swamp and remnants of tropical rain forest. The reserve covers an area of 150,000 hectares and is home to jaguars, spider monkeys, white-tail deer, peccaries, sea turtles and fishing eagles.

MONTES AZULES BIOSPHERE RESERVE, CHIAPAS

Montes Azules was established in southeast Chiapas in an attempt to save the last of the Lacandón Jungle. The reserve encompasses 331,200 hectares and is the largest in Mexico protecting tropical rain forest. Here huge buttressed mahoganies and chico zapotes reach a height of sixty meters. Hundreds of tree varieties, orchids, air plants, crocodiles, jaguars, tapirs and innu-merable insect species coexist here in an exuberance of life that defies accurate description.

EL OCOTE JUNGLE ECOLOGICAL RESERVE, CHIAPAS

El Ocote, located in the region of the northern moun-tains and encompassing 48,000 hectares, includes trop-ical evergreen and deciduous forests, providing refuge for pumas, spiker monkeys, peccaries, harpy eagles and river crocodiles.

CALAKMUL BIOSPHERE RESERVE, CAMPECHE

Calakmul protects a vast expanse of tropical evergreen forest in the south-central portion of the Yucatán peninsula and comprises the northern sector of the proposed Maya Peace Park. This proposal will unite it with the Maya Biosphere Reserve in Guatemala and the Rio Bravo Conservation Area in Belize to form the largest block of protected rain forest north of the Amazon Basin. This entire area is under assault by poachers and illegal lumbering operations, and only a major international effort will be effective in saving this pristine forest.

Sian Ka'an Biosphere Reserve, Quintana Roo

Just south of the ruins of Tulum, a dirt track—the main road to Punta Allen—runs past the entrance to Sian Ka'an. Established for the conservation of 528,000 hectares of beaches, mangrove swamps, estuaries and tropical evergreen forest, this reserve offers unspoiled tropical wilderness with excellent opportunities for trekking and camping.

Rio Lagartos Special Biosphere Reserve

Rio Lagartos, on the northern tip of the Yucatán peninsula, protects 47,840 hectares of dunes, beaches, mangrove swamps and dry thorn forest and is the best place to see pink flamingos.

Xel-Há National Park, Quintana Roo

Known as the world's largest natural aquarium, Xel-Há has become extremely popular with tourists. Visitors can rent masks for snorkeling and can paddle around in the largest fresh-water creek on the coast of Quintana Roo. The waters are teeming with tropical fish varieties of every size, shape and color.

Belize

Cockscomb Basin Wildlife Reserve, Stann Creek District

Scientists have recorded the highest density of jaguars in the world in this reserve of 41,450 hectares of low jungle. The chances of running into this largely nocturnal cat are small, and the reserve is an excellent place for camping, bird-watching and swimming in its pristine rivers and waterfalls.

Chiquibul National Park, Cayo District

Belize's newest and largest reserve protects 107,607 hectares of wilderness, including the southern portion of the Maya Mountains and the Vaca Plateau. Accessed by the road to the Mountain Pine Ridge out of San Ignacio Town, this wild tropical rain forest is suitable for visiting only for the experienced backwoods traveler.

Half Moon Cay National Park, Lighthouse Reef

Originally set aside for the conservation of a colony of several thousand red-footed booby birds in 1928, the reserve was expanded to include 4,144 hectares of habitat in 1982 when it gained national park status. Among the other inhabitants of the reserve are frigate birds and mangrove warblers as well as seventy-seven migratory bird species. Loggerhead and hawksbill sea turtles lay their eggs on the sandy eastern beaches of the Cay.

Hol Chan Marine Reserve

This 1,295-hectare underwater park is accessible only by boat. It protects a section of reef habitat located between Cay Cauker and Ambergris Cay, offering snorkelers and scuba divers the opportunity to see an array of sharks, manta rays, green moray eels and brilliantly colored coral and sea fans.

Rio Bravo Conservation Area, Orange Walk District

This large chunk of tropical evergreen and rain forest is home for jaguars, margays, crocodiles, howler and spider monkeys and 250 bird species. Rio Bravo is the second largest reserve in Belize (68,752 hectares) and is reached by dirt track from Orange Walk Town or by private plane.

Chan Chich Nature Reserve, Orange Walk District

Next door to the Rio Bravo Reserve lies Chan Chich, privately owned and managed by pioneer Belizean conservationist Barry Bowen. It is widely considered to be the best-protected and most well-managed reserve in Belize. Visitor accommodations at the Chan Chich Lodge include modern thatch-roofed cabins set around a Classic-period Mayan plaza and ruins. The reserve provides refuge to five species of wild cats, more than 250 bird species, monkeys, crocodiles, coatimundis, tapirs and kinkajous.

Blue Hole and St. Herman's Cave National Park, Cayo District

This small 233-hectare park is situated along the Hummingbird Highway, a short drive south of the Belizean capital of Belmopan. The Blue Hole is formed by an underground river that emerges as a spring and runs on the surface, only to disappear again fifty meters away into the mouth of another cave. Its cool turquoise waters are surrounded by the lush vegetation of the rain forest. St. Herman's Cave contains some beautiful formations.

COMMUNITY BABOON SANCTUARY, BELIZE DISTRICT

This private reserve is run by a cooperative of local residents and was established for the protection of the howler monkey (known in Belize as "baboon") and its rain forest habitat. It lies within and around the village of Burrel Boom and provides an excellent opportunity to observe wild howlers as they roam the forest canopy in troops of ten or more individuals.

SHIPSTERN NATURE RESERVE, COROZAL DISTRICT

Protecting three distinct types of tropical hardwood forests—including more than one hundred tree species—this 9,000-hectare reserve is located near the coastal village of Sarteneja. Some two hundred bird species, two hundred butterfly species and sixty species of reptiles and amphibians have been identified within the reserve, which also includes several saline lagoons and a shoreline covered with mangroves.

CROOKED TREE WILDLIFE SANCTUARY, BELIZE DISTRICT

Crooked Tree, just a short drive from Belize City on the Northern Highway, includes extensive swamps and marshlands and was set up especially to provide habitat for a growing number of endangered bird species. Of note among the reserve's residents is the Jabiru stork, the Western Hemisphere's largest bird, with a wing span of twelve feet. The reserve is also home to lesser-yellowlegs, social flycatchers and purple gallinules.

Guatemala

MAYA BIOSPHERE RESERVE, EL PETÉN PROVINCE

By far the largest reserve in Guatemala and one of the most ambitious projects in the world dedicated to protecting rain forest, the Maya Biosphere Reserve encompasses 13,000 square kilometers of tropical evergreen forest and tropical rain forest. It is part of the proposed Maya Peace Park (see Calakmul Biosphere Reserve, Mexico) and provides refuge to countless species, including the jaguar and the harpy eagle, the largest eagle on Earth. Within the reserve at least twenty-five major Mayan ceremonial centers have been located.

BIOTOPO CERRO CAHUÍ, EL PETÉN PROVINCE

Located along the northeast shore of Lake Petén Itzá, this small reserve (750 hectares) conserves a patch of rain forest covering Cahuí Hill. A well-maintained system of trails leads visitors up into the unspoiled forest. The reserve is a great place for bird watchers, it is easy to get to and it is located near a variety of accommodations.

BIOTOPO LAGUNA DE TIGRE-RIO ESCONDIDO, EL PETÉN PROVINCE

It is very difficult to get to this remote reserve, and with virtually no infrastructure, it is of interest only to the serious explorer. Located in the northwest sector of El Petén, this is one of the last places where one can see pairs of scarlet macaws flying over the rain forest canopy or meet up with prehistoric-looking tapirs and tamandua anteaters. The area is threatened by illegal logging operations and oil exploration.

LAGUNA DE LACHUÁ NATIONAL PARK, ALTA VERAPAZ PROVINCE

This reserve protects 10,000 hectares of rain forest, in the middle of which shimmer the placid blue waters of Lachuá Lagoon, a true jungle paradise. It is located near the frontier town of Playa Grande, and the easiest way to get there is to fly from Cobán to Playa Grande and then drive on dirt track to the park entrance.

LOS CUCHUMATANES NATIONAL PARK, HUEHUETENANGO PROVINCE

This alpine habitat in north-central Guatemala is reached by the all-weather road that links the provincial capital of Huehuetenango with the frontier town of Barrillas. The road passes through beautiful pine-oak forests with impressive cliffs and rock formations, and as it winds its way high up into the Cuchumatanes Mountains it emerges above the tree line into a fairyland of alpine meadows sprinkled with wildflowers and strange-looking agaves and succulents.

SIERRA DE LAS MINAS BIOSPHERE RESERVE; EL PROGRESO, ALTA AND BAJA VERAPAZ PROVINCES

Established to protect a wide diversity of habitats—from dry thorn forest to pine-oak forest and cloud forest—Sierra de las Minas is one of the wildest places on Earth and encompasses 105,700 hectares. Its extremely rugged terrain presents a natural barrier to those who would exploit its enormous reserves of lumber, so there is reason to hope that this wilderness will remain intact despite government corruption and insufficient conservation efforts. This reserve is only for the experienced hiker (a guide is recommended), and access trails begin from the towns of Chilasco and San Augustine.

Biotopo Mario Dary Rivera Quetzal Sanctuary, Baja Verapaz Province

The easiest way to reach the cloud forest home of the resplendent quetzal is via this reserve, known locally as Biotopo del Quetzal. Dedicated to the memory of its creator, pioneer Guatemalan conservationist Mario Dary, this small reserve, with some eight kilometers of well-maintained trails, is a must for anyone interested in tropical nature.

Lake Atitlán National Park, Sololá Province

Considered by many to be the most beautiful lake on Earth, Atitlán's sparkling, blue-green, clear waters are accentuated by the conical peaks of three extinct volcanoes that lie along her southern shore. Climbers who scale the volcanoes will find intact cloud forest. Ten villages populated by Cak'Chiquel and Tzutuhíl Indians are situated along the lake.

Río Dulce National Park, Izabal Province

The upper reaches of the "Sweet River" have been over-developed as a tourist destination, but downriver the jungle still reigns, and a wide variety of creatures, from howler monkeys to parrots and toucans, may be seen. As the river approaches the Caribbean Sea and the little Garifona town of Livingston, it meanders through a canyon with high vertical walls covered with lively vegetation and wildflowers.

Biotopo Chocon-Machacas Manatee Reserve, Izabal Province

Located along the northern edge of El Golfete, this reserve protects 6,265 hectares of rain forest as well as a series of lagoons and mangrove swamps. The reserve is an excellent place for bird-watching and for the study of tropical aquatic habitats, although the chances of sighting a manatee are slim.

Pacaya Volcano National Park, Guatemala Province

Groups leave daily from Antigua to visit the best place in the region to view an active volcano. After dark, spectacular eruptions are a common site.

Biotopo Monterrico-Hawaii, Santa Rosa Province

This beautiful reserve on Guatemala's south coast has been set aside to conserve mangrove swamp and estuary habitats and to protect the nest sites of several species of endangered sea turtles.

Honduras

El Trifinio National Park, Ocotepeque Province

The Honduras section of the Trifinio-La Fraternidad International Biosphere Reserve includes 5,400 hectares of pine-oak forest and cloud forest but lacks the excellent system of trails found in El Salvador's portion of the reserve, Montecristo National Park.

El Pital Biological Reserve, Ocotepeque Province

Located ten kilometers west of El Trifinio, this primitive reserve is also cloud forest and includes the third highest point in Honduras at an elevation of 2,730 meters above sea level.

Azul-Meambar National Park, Santa Barbara Province

An undeveloped park, Azul-Meambar protects a pine-oak forest running along the range of mountains east of Lake Yojoa.

Guisayote Biological Reserve, Ocotepeque Province

Guisayote protects thirty-five square kilometers of cloud forest and can be reached from the town of El Portillo. The park has a good system of trails, and with a little patience the visitor can see quetzals, mountain trogons, emerald toucanets and possibly an arboreal margay.

Celaque National Park, Copán Province

Access to Celaque is by way of the picuresque little town of Gracias, one of the country's oldest colonial settlements. It is a rugged day-hike up to the summit of the park (at 2,849 meters, the highest point in Honduras); the trail passes through pine forest and up into cloud forest at higher elevations.

Cusuco National Park, Cortéz Province

This reserve was established in 1951 when a Venezuelan ecologist reported that the area contained the largest pine trees in all of Central America. Cusuco includes areas of rain forest, pine and cloud forest and can be reached by way of abandoned logging roads. Two mountains stand within the park: Cerro Jilinco reaches 2,242 meters, and Cerro San Idelfonso, 2,228 meters.

La Tigra National Park, Francisco Morazán Province

La Tigra is a beautiful cloud forest reserve with an excellent system of trails that wind past waterfalls and lush groves of huge tree ferns. The highest point in the park is El Picacho (2,270 meters), and the park is only a short drive from Tegucigalpa, the capital of Honduras.

LANCETILLA BOTANCIAL GARDENS AND FOREST RESERVE

Lancetilla Botanical Gardens, located just outside the old banana port of Tela, was established in 1926 by the United Fruit Company as a plant research station. The beautiful botanical gardens have a large number of native species as well as the largest collection of exotic fruiting varieties in the Americas. The virgin rain forest that lies above the gardens was declared a reserve in 1990 and protects 1,681 hectares of prime tropical hardwood forest.

PUNTA SAL NATIONAL PARK, ATLANTIDA PROVINCE

Punta Sal is a peninsula jutting out into the Gulf of Honduras, a short distance from Tela. The reserve includes dunes and estuary and mangrove habitats and protects 41,250 hectares of oceanside habitat.

PICO BONITO NATIONAL PARK, ATLANTIDA PROVINCE

This wild and undeveloped park is located just outside the port town of La Ceiba. It rises radically from the coastal plain to an elevation of 2,433 meters, and a trek to the peak takes nine or ten days over tough terrain.

RIO PLATANO BIOSPHERE RESERVE, GRACIAS A DIOS PROVINCE

Honduras's largest nature reserve, in the far northeastern part of the country, was established by the United Nations and the Honduran government to preserve the outstanding biological diversity found in the Río Platano river valley. The immense tropical rain forest provides shelter to many endangered species, including the harpy eagle, jaguar, scarlet macaw and tapir. Numerous ruins lie within the reserve, and according to legend, the fabled lost White City of the Maya lies hidden under the jungle near the river's headwaters. Visitors can reach the reserve by either flying in or taking a boat from La Ceiba.

El Salvador

CERRO VERDE NATIONAL PARK, SONSONATE PROVINCE

Cerro Verde encompasses 6,500 hectares of mixed pine-oak forest and is the most visited park in El Salvador. A system of well-maintained trails takes visitors on a tour of the forest and affords excellent vistas of Izalco Volcano and Lake Coatepeque; visitors can also hike up to the summit of Izalco at nearly 1,910 meters. The site offers a hotel, restaurant and campgrounds.

DEININGER WILDLIFE RESERVE, LA LIBERTAD PROVINCE

This small 732-hectare reserve protects what is probably the last island of coastal tropical dry forest in El Salvador. It is situated a few miles southeast of the Port of La Libertad and includes a beautiful valley and a series of good trails. One path leads up to a rocky overlook, affording an excellent view of the park and a great vantage point for bird-watching.

EL IMPOSIBLE NATIONAL PARK, AHUACHAPAN PROVINCE

El Imposible was off-limits until recently because it was used as a guerilla base of operations. But a peace accord has now made it safe to visit the park, although the lack of any infrastructure limits access to all but experienced hikers. Located near the Guatemalan border, El Imposible's 5,600 hectares include examples of pine-oak-liquidambar and tropical evergreen forest.

MONTE CHRISTO NATIONAL PARK, SANTA ANA PROVINCE

This is the El Salvador portion of the Trifinio-La Fraternidad International Biosphere Reserve, which joins this forest with adjacent reserves in Honduras and Guatemala. Well-maintained with a visitor center, bunk houses and camping area, the park offers a trail leading up through the cloud forest to the 2,418-meter summit. A view tower at the top allows the visitor to take in a superb view of the surrounding forest.

NANCUCHINAME NATIONAL PARK, USULUTÁN PROVINCE

This undeveloped reserve created for the conservation of 1,175 hectares of coastal forest and mixed pine-oak forest is of special importance as a refuge for migratory bird species.

EL JOCOTAL WILDLIFE RESERVE, SAN MIGUEL PROVINCE

Surrounding Lake Jocotal, this small reserve protects habitat essential for the survival of a large number of birds, small mammals, reptiles and amphibians.

BARRA DE SANTIAGO WILDLIFE RESERVE, AHUACHAPAN PROVINCE

Barra de Santiago is the only site established for the conservation of mangrove swamp and marine estuary habitats in the country. The 2,200-hectare reserve in the southwestern sector of El Salvador is home to large numbers of shore birds as well as migratory bird species traveling north and south.

Contemporary Mayan Indian Settlements

Mexico

LACANJÁ, CHIAPAS

Considered to be the most traditional living Mayan tribe, the Lacandón Indians live in several small villages in what remains of the Lacandón Jungle of southeast Chiapas. They are thought to be the descendants of the people who built the ceremonial centers of Palenque, Yaxchilan and Bonampak. The Lacandón moved deep into the rain forest four hundred years ago in order to escape contact with the Spanish conquistadores, and during the ensuing centuries they continued to avoid the influences of Spanish colonization and Christianization. Only during the last couple of decades has this situation begun to change, as their jungle home has been invaded by land-hungry colonists, logging companies and oil exploration teams. Lacanjá is located on the way to the ruins of Bonampak and is the most accessible of the Lacandón villages.

These natives are instantly recognizable because the men still wear long hair and are clothed in simple, white tunics reaching down to their knees. By now they are used to visitors, and most of the men can speak some Spanish. An important part of their income comes from selling souvenirs to tourists in the form of carvings and miniature bow-and-arrow sets. With only around four hundred members in the whole tribe, it is very difficult for these people to resist the encroachment of modern technological society.

SAN JUAN CHAMULA, CHIAPAS

The Chamulas are Tzotzil-Maya, and their village is one of the most traditional in the area. Located near the colonial town of San Cristobal de las Casas in the central highlands of Chiapas, the village of Chamula is made up of 16,000 people, but only about three hundred of them live in the town itself while most are dispersed among little farms throughout the valley. This arrangement mirrors that of an ancient Mayan city, where large numbers of peasant farmers lived and worked the land surrounding the ceremonial center populated by priests and the upper class. The feudal cacique system—in which a few powerful individuals of the tribe own most of the land while the rest work as sharecroppers—still in force in Chamula is also similar to the social system common to Mayan cities of one thousand years ago.

In the middle of town, where in the past a pyramid might have stood, is the church. Inside the air is full of copal incense. There are no pews; the hall is an open expanse. Along the walls glass cases contain statues of the saints, each one with a mirror hanging around its neck. The Indians believe that when their images are reflected by these mirrors, a direct connection is established between the saints and themselves. The entire floor of the church is covered with sweet-smelling pine needles, and individuals and small groups of men, women and children sit or kneel on the floor. Small thin candles of red, yellow, green, white and black are stuck on the floor, each color having a special meaning. Bottles of *aguardiente* (cane liquor) and Coca Cola are placed by the candles to be blessed and later are drunk for the alleviation of spiritual or physical suffering.

Photography within the church is absolutely prohibited. Individuals who have ignored this rule have been severely beaten and their cameras destroyed.

TENEJAPA, CHIAPAS

The Sunday market at Tenejapa, a short drive from San Cristobal de las Casas, is one of the most interesting in the region. The women wear beautiful *huipils* (blouses) richly embroidered in bright colors, while the men are clothed in traditional black wool tunics and sport woven hats festooned with multihued ribbons. The inhabitants of the town are Tzeltal-Maya and are primarily dedicated to agriculture and weaving.

Belize

MAYA CENTER, STANN CREEK DISTRICT

The first time I visited Maya Center, a Mopan-Maya settlement in south-central Belize, I was impressed when a young native man approached me and, in perfect English, asked me if I needed any assistance. Because of the British educational system present in Belize, all young people attend school and receive an excellent primary and secondary education. This is in marked contrast to Guatemala, where most Indians never get the chance to attend school at all.

The Mopan Indians, although more sophisticated in the modern sense than other Mayan Indian groups, still maintain their language and many of their ancient customs in a country where they are not considered second-class citizens. They are treated with equality and respect.

Guatemala

TODOS SANTOS CUCHUMATÁN, HUEHUETENANGO PROVINCE

Todos Santos Cuchumatán is one of the most traditional and interesting Indian towns in all of Guatemala. High up in a valley within the boundaries of the proposed Cuchumatanes National Park, the town is home to about two thousand Mam Indians, with many more living in the surrounding hills. This is one of the few places left where Mayan day-keepers still use the ancient 260-day Tzolkin calendar. Some of the finest weavings are created here, and nearly all the residents still use their traditional costume; the men wear red-and-white striped trousers with wool breechcloths and striped shirts with huge, multicolored collars. A series of burial mounds from the pre-Classic ceremonial center of Teumanchún is located a few blocks from the town center. Todos Santos can be reached by a fifty-four–kilometer, badly rutted dirt track from the provincial capital of Huehuetenango.

ZUNIL, QUETZALTENANGO PROVINCE

In between the conical peaks of three volcanoes lies the little Indian town of Zunil. A Catholic church stands at the village center, and the plaza in front of the church is the site of Monday and Friday markets, where women dressed in pink, purple and red *huipils* sell fruits, vegetables and various wares. The church houses an ornate altar made of pure silver in front of which residents pray to both Christian and ancient Mayan deities.

The shrine of Maximón is behind the church in a two-story, green building, and on several occasions Catholic priests have made futile attempts to have him destroyed.

SANTIAGO ATITLÁN, SOLOLÁ PROVINCE

This traditional Tzutuhíl Indian village nestles at the foot of the towering Santiago Volcano, and its proud people have continued to resist European influence and maintain many of their ancient customs intact. The men cultivate corn, beans and squash and fish on Lake Atitlán; the women take charge of the household economy, nurture the children and weave the clothing. Santiago has just recently emerged from ten years of great suffering: the village was occupied by the army, and many Tzutuhíl were kidnapped, tortured and reported "disappeared." The unwelcome occupation ended when the army shot down twenty-eight unarmed villagers who had marched on the base to protest abuses. Rather than become involved in an almost certain bloodbath aftermath, the army decided to withdraw. Today a spirit of liveliness and gaiety pervades the village.

Another struggle (though not a violent one) is brewing in the town, between traditional Indian shamans and evangelical missionaries. The evangelists, in contrast to the religious tolerance of the Catholic missionaries and priests, unequivocally denounce all traditional religious practices as evil, but residents of Santiago still consult numerous shamans or brujos. The outcome of this power struggle cannot be predicted, but as the Tzutuhíl have successfully resisted all attempts to destroy their culture for more than four centuries, it is reasonable to assume they will continue to do so for some time to come.

CHICHICASTENANGO, EL QUICHÉ PROVINCE

The largest traditional markets and some of the best weaving in all of the land of the Maya can be found in "Chichi." It is the most famous of all highland Indian towns and plays host to large numbers of tourist who flood in to see the Thursday and Sunday markets. The market is full of every imaginable commodity, from live turkeys and piglets to beautiful weavings, ceremonial masks, pots and pans, fruits and vegetables, cassette players and tapes, machetes and farm implements, copal incense and religious objects.

Here the priest Francisco Ximenez became the first outsider to be shown a manuscript of the *Popol Vuh*, the sacred book of the Quichés and the finest example of pre-Columbian literature to survive the conquest. It is said that when the Indians saw Ximenez show a respectful interest in the *Popol Vuh*, they decided to move some of their altars to the Catholic church, thereby beginning the melding of faiths that continues today.

NEBAJ, EL QUICHÉ PROVINCE

Nebaj is the largest of the three Ixíl Indian towns that comprise the Ixíl Triangle, the others being Cotzal and Chajul. This area was considered off-limits to travelers until recently because some of the worst war atrocities occurred here during the Guatemalan army's infamous scorched-earth campaign to rout guerrilla sympathizers. Today the number of tourists visiting the Ixíl Triangle is steadily increasing.

The settlements in the Ixíl Triangle are among the most remote and conservative of all Mayan villages. Situated on the slopes of the Cuchumatanes Mountains and accessible by driving several hours north from Chichicastenango, Nebaj's Indians have held steadfastly to their traditional community structures, customs and religious practices.

San Juan Cha Melco, Alta Verapaz Province

This little Kek'chi Indian town is probably not listed in any guide book because nothing in the village stands out as spectacular or of unusual interest to tourists. Nevertheless, it provides a prime opportunity for learning about the Kek'chi people and culture. Located a short drive from the provincial capital of Cobán, Chamelco is home to several thousand friendly and hard-working souls. Most men no longer wear traditional costumes, but the women are still decked out in fine *cortes* (skirts) and exquisitely embroidered *huipils.* Agriculture is the mainstay of this village, as it has always been. The hardy traveler can make an excursion up into the Sierra Yalihúx cloud forest to the little hamlet of Chelemja, so remote that its inhabitants do not speak Spanish.

Recommended Reading

Alvarez Del Toro, M. 1978. *Mamiferos de Chiapas.* Tuxtla Gutiérrez: Instituto de Historia Natural.

———. 1980. *Aves de Chiapas.* Chiapas: Universidad Autonoma de Chiapas.

———. 1982. *Los Reptiles de Chiapas.* Tuxtla Gutiérrez: Instituto de Historia Natural.

Alvarez Del Toro, M. et al. 1993. *Chiapas y su Biodiversidad.* Chiapas: Gobierno de Chiapas.

Barry, T. 1990. *Guatemala: A Country Profile.* Albuquerque, N. Mex.: Interhemispheric Education Resource Center.

———. 1992. *Mexico: A Country Guide.* Albuquerque, N. Mex.: Interhemispheric Education Resource Center.

Campbell, J. A., and W. W. Lamar. 1992. *The Venomous Snakes of Latin America.* Ithaca, N.Y.: Cornell University Press.

Carmack, R. M., ed. 1992. *Guatemala: Harvest of Violence.* Norman, Okla.: University of Oklahoma Press.

Coe, M. D. 1966. *The Maya.* London: Thames and Hudson.

———. 1992. *Breaking the Maya Code.* New York: Thames and Hudson.

Diaz-Bolio, J. 1987. *The Geometry of the Maya.* Merida, Mexico: Mayan Area Publications.

Edmonson, M. S., trans. 1982. *The Ancient Future of the Itzá.* Austin, Tex.: University of Texas Press.

Harris, H., and S. Ritz. 1993. *The Maya Route.* Berkeley, Calif.: Ulysses Press.

Holdridge, C. 1967. *Life Zone Ecology.* San Jose, Costa Rica: Tropical Science Center Publications.

Hunter, B. C. 1986. *A Guide to the Ancient Maya Ruins.* Norman, Okla.: University of Oklahoma Press.

Janson, O. T. 1981. *Animales de Centroamerica en Peligro.* Guatemala City: Editorial Piedra Santa.

———. 1992. *Quetzal.* Guatemala City: Editorial Artemes y Edinter.

Kricher, J. C. 1989. *A Neotropical Companion.* Princeton, N.J.: Princeton University Press.

Land, H. C. 1970. *Birds of Guatemala.* Philadelphia: Livingston Publishing Co.

Leal, M. C. 1991. *Archaeological Mexico.* Milano: Casa Editrice Bonechi.

Leopold, A. S. 1959. *Wildlife of Mexico.* Berkeley, Calif.: University of California Press.

Mahler, R. 1993. *Guatemala: A Natural Destination.* Santa Fe, N. Mex.: John Muir Press.

Mallan, C. 1993. *Belize Handbook.* Chico, Calif.: Moon Publications.

———. 1993. *Yucatán Peninsula Handbook.* Chico, Calif.: Moon Publications.

Miller, O. 1976. *The Day Spring.* Toronto: McClelland and Stewart Limited.

Peterson, R. T., and E. L. Chalif. 1973. *A Field Guide to Mexican Birds.* Boston: Houghton Mifflin Co.

Schele, L., and D. Freidel. 1990. *A Forest of Kings.* New York: William Morrow.

Simon, J. 1987. *Guatemala: Eternal Spring, Eternal Tyranny.* New York: W. W. Norton.

Stephens, J. L. 1969. *Incidents of Travel in Central America: Chiapas and Yucatán.* New York: Dover Publishers.

Tedlock, D. 1985. *Popol Vuh.* New York: Simon and Schuster.

Tompkins, P. 1976. *Mysteries of the Mexican Pyramids.* New York: Harper and Row.

Ugalde, A., and J. C. Godoy. 1992. *Areas Protegidas de Centroamerica.* Gland, Switzerland: International Union for the Conservation of Nature and Natural Resources.

Wright, R. 1989. *Time Among the Maya*. New York: Weidenfeld & Nicholson.

Central American Handbook. 1994. New York: Prentice Hall.

Mexico and Central America Handbook. 1993. New York: Prentice Hall.

About the Authors

THOR JANSON has worked in Central America for nearly twenty years in a variety of capacities, including biologist, conservationist, writer, photographer and entrepreneur. In 1976 he was invited by Mario Dary, Guatemala's pioneer conservationist, to join the faculty at San Carlos University. There Janson designed and supervised a project to study the endangered manatee, which led to the establishment of a large protected reserve on the Caribbean side of the country. In 1981 Dary, by this time the rector at San Carlos, was assassinated, and Janson decided to leave the politically volatile university and create his own conservation organization. In 1982 he founded Defensores de la Naturaleza (Defenders of Nature) for the purposes of developing environmental education campaigns and establishing a system of privately funded and managed reserves in Guatemala. He worked as director of Defensores for five years, during which time the group became influential in promoting the preservation of rain forests. At present Janson is executive director of Embajada del Reino Natural. He is the author of five books and more than one hundred articles concerning nature and the Mayan world.

ROBERT M. CARMACK is professor of anthrolopology at the State University of New York at Albany. He received his Ph.D. in social anthropology from the University of California, Los Angeles, in 1964. Carmack's field work in Mesoamerica has been carried out over more than twenty-five years, during which time he has spent more than fifty months living among the Maya. He has published seven books and thirty articles on his research findings in Guatemala.